Studies in Systems, Decision and Control

Volume 502

The series "Studies in Systems, Decision and Control" (SSDC) covers both new developments and advances, as well as the state of the art, in the various areas of broadly perceived systems, decision making and control–quickly, up to date and with a high quality. The intent is to cover the theory, applications, and perspectives on the state of the art and future developments relevant to systems, decision making, control, complex processes and related areas, as embedded in the fields of engineering, computer science, physics, economics, social and life sciences, as well as the paradigms and methodologies behind them. The series contains monographs, textbooks, lecture notes and edited volumes in systems, decision making and control spanning the areas of Cyber-Physical Systems, Autonomous Systems, Sensor Networks, Control Systems, Energy Systems, Automotive Systems, Biological Systems, Vehicular Networking and Connected Vehicles, Aerospace Systems, Automation, Manufacturing, Smart Grids, Nonlinear Systems, Power Systems, Robotics, Social Systems, Economic Systems and other. Of particular value to both the contributors and the readership are the short publication timeframe and the world-wide distribution and exposure which enable both a wide and rapid dissemination of research output.

Indexed by SCOPUS, DBLP, WTI Frankfurt eG, zbMATH, SCImago.

All books published in the series are submitted for consideration in Web of Science.

Predrag Petrović

Processing, Estimation and Measurement of Signals Parameters in Public Distribution Networks

Springer

Predrag Petrović
Faculty of Technical Sciences Čačak
University of Kragujevac
Čačak, Serbia

ISSN 2198-4182 ISSN 2198-4190 (electronic)
Studies in Systems, Decision and Control
ISBN 978-3-031-43106-7 ISBN 978-3-031-43107-4 (eBook)
https://doi.org/10.1007/978-3-031-43107-4

This Springer imprint is published by the registered company Springer Nature Switzerland AG
The registered company address is: Gewerbestrasse 11, 6330 Cham, Switzerland

Paper in this product is recyclable.

Preface

This monograph is the result of a long-term work of the author on the problems of digital processing of complex periodic signals of voltage and current which can be found in the distribution network. This scientific field has been given special attention in the literature abroad through a great number of articles published in the leading journals, textbooks and the other publishing forms.

Over the recent period, the author of the monograph have published a large number of papers in a number of journals and have presented at the leading international conferences, which verifies the results they have achieved in this scientific field.

The problem of the reconstruction and estimation of complex periodic signals, which is in the focus of Chapter three of this monograph, has been given special attention. A completely new protocol, which enables the development of much superior and more efficacious algorithms, has been developed, and the obtained results are unique in global practice.

In other Chapters processing of parameters of ac signals, is performed on basis of Hilbert, Newton-Raphson, Gramm matrix, Taylor-Fourier techniques. The conditions, required for the performance of such a processing, have necessarily been derived. Chapter six proposes a newly developed instrument for the calculation of basic parameters of the processed voltage and current signals. Having being practically verified, this method has demonstrated exceptionally favorable performance.

The authors believe that the monograph will be beneficial to specialists and experts involved in problems of this scientific field, and hope that it will serve as a useful guideline to follow in further investigations.

Čačak, Serbia | Predrag Petrović

Contents

Chapter 1
New Measurement Procedures Based on Measurements on Time Interval

In recent years, there has been a noticeable increase in the usage of nonlinear and time-varying electrical loads in common power systems. This can be attributed to the widespread adoption of power-electronic equipment, such as adjustable-speed drives, controlled rectifiers, cycloconverters, electronically ballasted lamps, arc and induction furnaces, and clusters of personal computers in industrial, commercial, and residential settings. However, these loads can lead to various disturbances affecting both the utility and the equipment of other customers. Consequently, the quality of electric power has become a significant concern for utilities and their customers, prompting them to adopt the guidelines and limits outlined in the new International Standards (e.g., IEC, EN, BS, IEEE).

The degradation of electrical power quality in commercial and industrial installations is undeniable. Apart from external disturbances like outages, sags, and spikes due to switching and atmospheric phenomena, inherent internal problems arise from the combined use of linear and non-linear loads at each site. Given these challenges, it is essential to develop measuring systems capable of accurately calculating basic AC values, considering the presence of higher harmonic components within the processed voltage and current signals [1–4].

The IEEE Standard 1459-2000 defines power components such as active, reactive, distortion, nonactive, and apparent powers using fast Fourier transforms (FFT). While FFT can provide accurate results for stationary waveforms, it introduces significant errors for nonstationary waveforms due to spectral leakage and picket fence phenomena [5]. In contrast, the wavelet transform, a time–frequency transform, has proven effective in accurately representing nonstationary waveforms in various applications across different disciplines [6, 7]. The wavelet transform offers variable frequency resolution while preserving time information, making it suitable for analyzing and measuring nonstationary waveforms with time-variant characteristics, which are lost when using the FFT.

P. Petrović, *Processing, Estimation and Measurement of Signals Parameters in Public Distribution Networks*, Studies in Systems, Decision and Control 502,
https://doi.org/10.1007/978-3-031-43107-4_1

Traditional algorithms for measuring active power and effective values of voltage and current rely on integration or summation over a defined time interval [8–14]. Synchronous sampling of alternating current (AC) signals enables highly accurate recalculation of basic electric values in a network, particularly when dealing with spectrally limited signals and having sufficient processing time and calculation capacities. However, this method requires precise measurement of the signal period T and generation of the sampling interval $T_S = T/N$, where T is the period of the processed signal, and N is the number of measurements necessary for exact calculation [8, 9]. It is suitable for sinusoidal and complex-periodical signals with low harmonic content.

Some methods [12–14] offer exceptional precision in calculating basic electrical values but require a complex calibration procedure and substantial processor power, making them suitable only for periodic input signals. The voltage and current waveforms in power networks, however, are not strictly periodic due to their nonharmonic components and stochastic variation. As a result, to maintain the high performance of such algorithms, it is necessary to analyze potential sources of errors in determining the integration interval [15] and introduce special estimation procedures [16–20], which add complexity to their structure. Consequently, the outputs of these algorithms often represent the input observed over a specific period rather than its instantaneous value. By utilizing estimated signal parameters [16–20], precise measurements of RMS value and other essential electrical quantities like power and energy can be achieved.

We present a novel approach for processing analog signals, especially AC signals in public distribution networks that can be represented using Fourier series. Our method achieves processing within a significantly reduced time interval, approximately half the duration compared to conventional techniques in current use [21, 22]. The primary focus of this approach is to determine the active power value or the effective value (RMS) of the processed signal without the need for synchronization. Notably, our method eliminates the requirement for the sampling frequency f_s of the analog-to-digital converter to be an integer multiple of the signal frequency f. This flexibility enhances the applicability of our approach in practical scenarios. Moreover, its implementation does not rely on complex hardware or computationally intensive numerical calculations, which contributes to its cost-effectiveness and simplicity. There is no necessity for any specialized estimation procedure to refine the obtained measuring results further.

We consider the system voltage $v_{input}(t)$ and current $i_{input}(t)$ signals as sums of their respective Fourier components, which can be expressed as:

$$v_{input} = \sqrt{2}V_R \cdot \sum_{r=1}^{M_1} k_r \sin(r\omega t + \psi_r) = v_{input}(t),$$
$$i_{input} = \sqrt{2}I_R \cdot \sum_{s=1}^{M_2} l_s \sin(s\omega t + \varphi_s) = i_{input}(t) \tag{1.1}$$

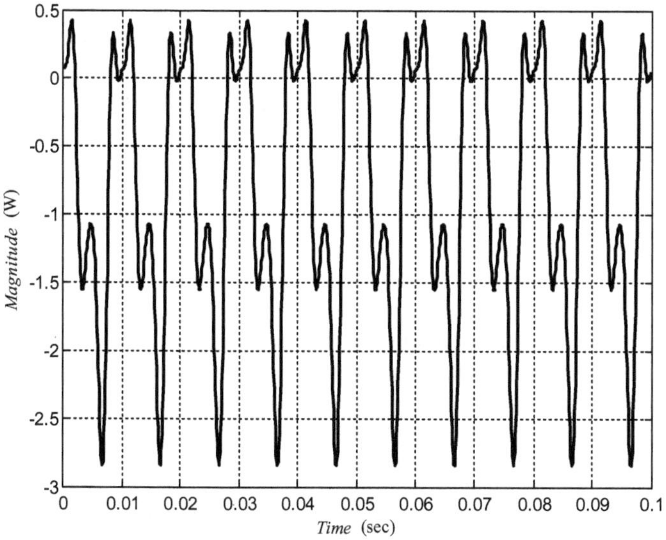

Fig. 1.1 The power signal obtained from the voltage signal that possesses 1 and 3 harmonics, and the current signal with 1, 3, 5, 7 and 9 harmonics

Let $\omega = 2\pi f$ denote the angular frequency, where f represents the frequency. $k_r V_R$ corresponds to the RMS voltage value of the rth harmonic, and $l_s I_R$ corresponds to the RMS current value of the sth harmonic. Additionally, ψ_r and ϕ_s represent the phase angles of the rth and sth harmonic components of voltage and current, respectively. M_1 and M_2 denote the numbers of the highest harmonic components of the voltage and current signals. Utilizing the known spectral limits, the frequency of the selection of processed signals can be determined in accordance with the Nyquist criteria. In real-world scenarios, the current signal typically exhibits a more complex harmonic content.

The average power P is defined as follows:

$$P = \frac{1}{T}\int_0^T i_{input}(t)v_{input}(t)dt = \frac{1}{T}\int_0^T p(t)dt;\ T = \frac{1}{f} \tag{1.2}$$

Nevertheless, Fig. 1.1 [23] illustrates that the actual power signal of the band-limited signals exhibits periodicity. This periodic behavior allows us to adjust the size of the time interval used for calculating the active power value.

Through a more in-depth examination of Eq. (1.2), we arrive at the following result:

$$
\begin{aligned}
P &= \frac{1}{T_x}\int_{t_1}^{T_x+t_1}\left[\sqrt{2}I_R\cdot\sum_{s=1}^{M_2} l_s\sin(s\omega t+\varphi_s)\right]\cdot\left[\sqrt{2}V_R\cdot\sum_{r=1}^{M_1} k_r\sin(r\omega t+\psi_r)\right]dt \\
&= \frac{1}{T_x}\int_{t_1}^{T_x+t_1} 2I_R V_R\cdot\sum_{r=1}^{M_1}\sum_{s=1}^{M_2} k_r l_s\sin(r\omega t+\psi_r)\sin(s\omega t+\varphi_s)dt \\
&= \frac{1}{T_x} I_R V_R\cdot \\
&\cdot\sum_{r=1}^{M_1}\sum_{s=1}^{M_2} k_r l_s\int_{t_1}^{T_x+t_1}\{\cos[(r-s)\omega t+\psi_r-\varphi_s]-\cos[(r+s)\omega t+\psi_r+\varphi_s]\}dt
\end{aligned}
\tag{1.3}
$$

Equation (1.3) introduces the interval denoted as T_x, which serves as the foundation for calculating the active power using our proposed method. The specific value of T_x will be determined based on the analysis of the observed relation. To simplify Eq. (1.3), we introduce the following abbreviations:

$$
\begin{aligned}
A &= \frac{1}{T_x} I_R V_R\cdot\sum_{r=1}^{M_1}\sum_{s=1}^{M_2}{}_{r\neq s} k_r l_s\cdot\frac{1}{(r-s)\omega}\sin[(r-s)\omega t+\psi_r-\varphi_s]\Big|_{t_1}^{T_x+t_1} \\
B &= \frac{1}{T_x} I_R V_R\cdot\sum_{r=1}^{Q=\min(M_1,M_2)} k_r l_r\int_{t_1}^{T_x+t_1}\cos(\psi_r-\varphi_r)dt = I_R V_R\cdot \\
&\cdot\sum_{r=1}^{Q=\min(M_1,M_2)} k_r l_r\cos(\psi_r-\varphi_r) \\
C &= -\frac{1}{T_x} I_R V_R\cdot\sum_{r=1}^{M_1}\sum_{s=1}^{M_2} k_r l_s\cdot\frac{1}{(r+s)\omega}\sin[(r+s)\omega t+\psi_r+\varphi_s]\Big|_{t_1}^{T_x+t_1}
\end{aligned}
\tag{1.4}
$$

The variable B in Eq. (1.4) represents the precise value of the active power of the processed voltage and current signal, as defined by relation (1.1) following standard [24]. To ensure an accurate calculation of the active power, it is crucial for the variables A and C in Eq. (1.4) to be equal to 0 [9]. This condition dictates the necessary value of the interval T_x, which serves as the basis for computing the active power. Expanding variable A (and similarly, C) in the aforementioned equation yields the following outcome:

$$
A = \frac{1}{T_x} I_R V_R\cdot\sum_{r=1}^{M}\sum_{s=1}^{M}{}_{r\neq s} k_r l_s\cdot\frac{1}{(r+s)\omega}\left\{
\begin{array}{l}
\sin[(r-s)\omega T_x]\cos\left[(r-s)\omega t_1\right]\cos(\psi_r-\phi_s)+ \\
+\cos[(r-s)\omega T_x]\sin\left[(r-s)\omega t_1\right]\cos(\psi_r-\phi_s)+ \\
+\cos[(r-s)\omega T]\cos\left[(r-s)\omega t_1\right]\sin(\psi_r-\phi_s)- \\
-\sin[(r-s)\omega T]\sin\left[(r-s)\omega t_1\right]\sin(\psi_r-\phi_s)- \\
-\sin\left[(r-s)\omega t_1\right]\cos(\psi_r-\phi_s)-\cos\left[(r-s)\omega t_1\right]\sin(\psi_r-\phi_s)
\end{array}
\right\}
\tag{1.5}
$$

Assuming that the voltage and current signals contain only odd harmonic components (as in public distribution networks [23, 25]), i.e., the indices r and s take the form $2\boldsymbol{k}+1$ in Eqs. (1.4) and (1.5), the integrals can be computed over the range from $\boldsymbol{t_1}$ to $\boldsymbol{T}/2+\boldsymbol{t_1}$ (equivalently, from 0 to $\boldsymbol{T}/2$), with variables A and C set to 0. Consequently, under this assumption, all terms of the form $r-s$ present in Eq. (1.5) must have even values, i.e., they will be in the form of $2m$ (where m is a whole number), leading to a sum of trigonometric terms in Eq. (1.5) that evaluates to 0.

This modification allows the calculation to be performed within a time interval that is twice shorter, i.e., using half the number of samples during digital processing, compared to the classical method of synchronous sampling. In a previous paper [26], the active power is computed in the semi-period of the processed signals by shifting the average filter, but no specific condition is derived in the time-domain to enable this kind of calculation, and an algorithm for re-calculation is not provided, as is done in this paper. Furthermore, the issue of synchronization with the processed signals, a challenge faced in the approach suggested in [26], has been successfully resolved in this current approach.

1.1 Accurate Active Power Calculation

The algorithm implementing Eq. (1.3) is outlined as follows:

(1) Sample the voltage signal V_k and the current signal I_k at a sampling frequency f_S.
(2) Detect voltage zero-crossing $\{V_{-1}, V_0\}$ and $\{V_{N-1}, V_N\}$ when:

$$V_K \cdot V_{K-1} < 0 \tag{1.6}$$

The determination of the number of zero-crossings in the voltage signal, as defined in Eq. (1.6), employs a sign-test on consecutive samples. This approach differs significantly from the methods discussed in [25, 27, 28], which involve more complex techniques such as phase-locked loops (PLL), harmonic filtering before or after sampling, or comprehensive signal filtering and post-processing. In contrast, the threshold crossing technique proposed in this paper offers a simpler alternative.

To enhance the resistance to noise, a window of $2n$-length data can be utilized. A zero-crossing (ZC) is considered valid and included in the window only when n consecutive samples exhibit positive signs followed by n consecutive samples with negative signs. By choosing an appropriate value for n, such as 3 to 5, the impact of random noise on erroneous ZC detection can be effectively minimized. Once a ZC is detected, its position along the time-axis can be determined using linear interpolation between two consecutive samples with different signs. Assuming the two samples are $V_k > 0$ and $V_{k-1} < 0$, the time-axis location of the ZC is calculated as follows:

$$x = k - \frac{V_K}{V_K - V_{K-1}} \tag{1.7}$$

Once the positions of two consecutive zero-crossings (ZC) are determined at x_1 and x_2, the frequency can be computed using the following formula:

$$f = \frac{1}{2T_s(x_1 - x_2)} \tag{1.8}$$

where $T_S = 1/f_S$ represents the sampling interval. For online frequency estimation, a sliding window of samples is utilized, with x_1 and x_2 referring to the start point of the data window. To ensure accuracy and exclude erroneous zero-crossings, a security condition is implemented. If the interval between two zero-crossings deviates significantly from the previous interval, the newly detected zero-crossing is considered invalid, and the previous frequency result is retained. This straightforward logic already provides satisfactory frequency accuracy. Accurately determining the frequency of processed AC signals is a significant challenge [4, 29]. Existing methods often involve complex mathematical techniques and extensive processing time. In contrast, the authors aimed to develop a simpler algorithm with lower numerical complexity while still achieving precise frequency estimation. The effectiveness of this approach was validated through extensive testing. The subsequent steps of the algorithm are as follows:

(3) Construct an array of power samples $\{P_0, P_1, ..., P_{N-1}\}$ between adjacent voltage zero-crossings.
(4) Calculate the active power using Eq. (1.3).
(5) Repeat steps (1)–(4) for the subsequent voltage half-periods.

The integrals in Eq. (1.3) are computed using the trapezium method [30], where the function $p(t)$ is approximated by piecewise linear segments between the samples, P_i. By performing the calculation of active power over the interval $[0, T/2]$, the following expression is obtained:

$$P' = \frac{2}{T}\frac{1}{f_S}\sum_{i=0}^{N-1} K_i P_i = \frac{2}{T}\frac{1}{f_S}\sum_{i=0}^{N-1} K_i V_i I_i \tag{1.9}$$

Let N denote the total number of samples, where K_i represents the coefficients derived from the trapezium method applied to the trapezium surfaces formed by consecutive instantaneous power samples.

$$\begin{aligned}
&K_1 = K_2 = ... = K_{N-2} = 1 \\
&\Delta T_0 = \frac{V_0}{V_0 - V_{-1}}\frac{1}{f_S}; \ \Delta T_{N-1} = \frac{V_N}{V_N - V_{N-1}}\frac{1}{f_S}; \\
&T_P = (N-1)\frac{1}{f_S} + \Delta T_0 + \Delta T_{N-1} \\
&K_0 = \frac{1}{2}(1 + \Delta T_0 \cdot f_S); \ K_{N-1} = \frac{1}{2}(1 + \Delta T_{N-1} \cdot f_S)
\end{aligned} \tag{1.10}$$

where ΔT_0 represents the time interval between the voltage crossing zero and the corresponding sample P_0, ΔT_{N-1} represents the time interval between sample P_{N-1} and the next voltage zero-crossing, and T_P represents the detected period of the instantaneous power ($T_p = T/2$). The selection of the coefficient K_i and the methodology employed in Eq. (1.8) introduces a calculation error. This error exhibits an opposing effect compared to the division error and is contingent upon the shape of the function $p(t)$. According to Eq. (1.9), the proposed algorithm necessitates N multiplications and N additions, thereby affirming its low computational complexity. As the integral is evaluated using the trapezium method, there exists an absolute error dependent on the integration time interval, sampling frequency, and the shape of the integrated function $p(t)$.

$$\begin{aligned} R(p(t)) &\approx -\frac{1}{12}\frac{T}{2}\left(\frac{1}{fs}\right)^2\frac{d^2}{dt^2}p(t) \\ &= -\frac{1}{12}\frac{T}{2}\left(\frac{1}{fs}\right)^2 I_R V_R \cdot \sum_{r=1}^{M_1}\sum_{s=1}^{M_2} k_r l_s \frac{d^2}{dt^2}\left\{\begin{array}{l}\cos[(r-s)\omega t+\psi_r-\varphi_s]- \\ -\cos[(r+s)\omega t+\psi_r+\varphi_s]\end{array}\right\} \\ &= \frac{1}{12}\frac{T}{2}\left(\frac{1}{fs}\right)^2 I_R V_R \omega^2 \cdot \sum_{r=1}^{M_1}\sum_{s=1}^{M_2} k_r l_s \left\{\begin{array}{l}(r-s)^2\cos[(r-s)\omega t+\psi_r-\varphi_s]- \\ -(r+s)^2\cos[(r+s)\omega t+\psi_r+\varphi_s]\end{array}\right\} \end{aligned} \quad (1.11)$$

The utilization of Eq. (1.9) for calculating active power introduces an inherent absolute error.

$$P' - P \approx -\frac{1}{12}\left(\frac{1}{fs}\right)^2 \frac{2}{T}\int_0^{T/2}\frac{d^2}{dt^2}p(t)dt \quad (1.12)$$

Equation (1.12) reveals that the estimated error is contingent upon the sampling frequency and signal period. The proposed method demonstrates immunity to signal-frequency variations at high sampling frequencies. However, at low sampling frequencies (fs), the error associated with the trapezium method becomes highly dependent on the signal frequency (f).

To minimize synchronization errors, it is advisable to select a sampling frequency (fs) that is not an integer multiple of the signal frequency (f). This choice ensures that the average jitter in the calculation of integral boundaries approaches zero. Notably, the synchronization error is negligible compared to the trapezium method error.

To simplify implementation, setting $t_1 = 0$ in Eq. (1.3) is most suitable. When the sampling frequency is sufficiently high, the outermost regions of the integral near the boundaries contribute insignificantly due to their small size. Thus, their accurate calculation is unnecessary, and the computational effort required by the proposed method (Eq. 1.9) can be reduced by increasing the sampling frequency. Consequently, the proposed method is well-suited for hardware realization in integrated technology, such as an IC power meter, thanks to its straightforward computational process. Moreover, synchronization is not required.

To calculate the RMS value of the input signal, it is necessary to square each sample and subsequently calculate the square root of the mean of the squared values. This formula defines the RMS signal value. The time required to obtain the desired value (RMS, active power) is slightly longer than half the period of the processed signal.

In a conventional energy system, this duration is slightly longer than 10 ms (or 8.33 ms for a 60 Hz network), which is significantly shorter than with alternative methods [31, 32]. When the RMS value is obtained through digital processing of a sequence of signal samples, the uncertainty and bias of the measured value depend on the chosen algorithm. As signal sampling is typically non-coherent in practice, leakage occurs in the signal's DFT spectrum, violating the definition of RMS for periodic signals in the time domain. The proposed algorithm overcomes these limitations, ensuring a high level of calculation precision.

Enhanced adaptability of the proposed algorithm can be achieved by increasing the number of samples of the processed voltage and current signals (by raising the sampling frequency, f_S) in a continuous time sequence from one measurement to another. This approach enables improved precision in calculating active power. However, it also introduces additional complexity to the practical implementation, which the authors aimed to avoid in the proposed solution. Another possibility is to reference the collection of voltage and current samples to a random moment instead of fixing it at $t_1 = 0$, as done in the proposed algorithm. Consequently, processing can occur within a time interval relative to the detected zero-crossings of the voltage signal. Although this alternative provides an additional degree of freedom, it also increases the complexity of the proposed solution. Thus, the final decision was made to exclude this modification.

1.2 Estimation of Measuring Uncertainty

The measurement uncertainty of the proposed algorithm serves as a critical metrological characteristic. To attain a highly precise measurement with minimal uncertainty, the entire decision-making process involved in conceiving, developing, and implementing the proposed method adhered to stringent requirements. A model representing the measurement results can be formulated by utilizing expressions (1.9) and (1.10).

$$\begin{aligned} P' &= \frac{2}{T}\frac{1}{f_s}\sum_{i=0}^{N-1} K_i P_i = \frac{2}{T}\frac{1}{f_s}\sum_{i=0}^{N-1} K_i V_i I_i = \\ &\frac{2}{T}\frac{1}{f_s}\left\{\frac{1}{2}\left(1+\frac{V_0}{V_0-V_{-1}}\right)V_0 I_0 + \sum_{i=1}^{N-2} V_i I_i + \frac{1}{2}\left(1+\frac{V_N}{V_N-V_{N-1}}\right)V_{N-1} I_{N-2}\right\} \end{aligned} \tag{1.13}$$

The coefficients $\frac{V_0}{V_0-V_{-1}}$ and $\frac{V_N}{V_N-V_{N-1}}$, with values ranging from 0 to 1, play a crucial role. When evaluating the measurement uncertainty of the active power P', it is reasonable to assume a fixed value for these coefficients. Specifically, setting them to 1 simplifies the expression for P', resulting in the following form:

$$P' \approx \frac{2}{T}\frac{1}{f_s}\sum_{i=0}^{N-1} V_i I_i \tag{1.14}$$

In accordance with the general expression for the standard measurement uncertainty (GUM) [33], the following is implemented:

$$
\begin{aligned}
&u^2(P') = \left[\tfrac{\partial P'}{\partial T}u(T)\right]^2 + \left[\tfrac{\partial P'}{\partial f_s}u(f_s)\right]^2 + \\
&\sum_{i=0}^{N-1}\left[\tfrac{\partial P'}{\partial V_i}u(V_i)\right]^2 + \sum_{i=0}^{N-1}\left[\tfrac{\partial P'}{\partial I_i}u(I_i)\right]^2 = \\
&\left(-\tfrac{1}{T}P'\right)^2 u^2(T) + \left(-\tfrac{1}{f_s}P'\right)^2 u^2(f_s) + \\
&\sum_{i=0}^{N-1}\left(\tfrac{2}{T}\tfrac{1}{f_s}I_i\right)^2\left(\tfrac{V_{range}}{2^{b+1}}\tfrac{1}{\sqrt{3}}\right)^2 + \sum_{i=0}^{N-1}\left(\tfrac{2}{T}\tfrac{1}{f_s}V_i\right)^2\left(\tfrac{I_{range}}{2^{b+1}}\tfrac{1}{\sqrt{3}}\right)^2
\end{aligned}
\tag{1.15}
$$

In accordance with the proposed algorithm, the measurement uncertainties of the values T, f_s, V_i and I_i contribute to the measurement uncertainty of the value P'. These values, $u(T)$ and $u(f_s)$, represent the standard measurement uncertainties (Type B) of the T and f_s values.

When simultaneous sampling of the voltage and current signals (the values V_i and I_i) is performed using b-bit AD converters, the assumption is made that the converters are ideal and that the only source of uncertainty is their limited resolution. In this case, the standard measurement uncertainty of the V_i value is a constant represented by $\frac{V_{range}}{2^{b+1}}\frac{1}{\sqrt{3}}$ (type B uncertainty), where V_{range} equals the measurement range of the applied AD converter. The same approach is employed to determine the standard measurement uncertainty of the current I_i values. However, in real-world scenarios, it is necessary to consider various factors that impact the precision of the AD converter, leading to an increase in the error limits to $\frac{V_{range}}{2^{b-1}}$ (for instance, an 11-bit precision for a 12-bit AD converter). It is not uncommon in practice for the precision of an AD converter to be specified as, for example, 14 bits for a 16-bit converter. In such cases, the standard measurement uncertainty of the voltage samples would amount to $\frac{V_{range}}{2^{b-2}}\frac{1}{\sqrt{3}}$. The expression for the measurement uncertainty (assuming ideal converters) can be expressed as follows:

$$
\begin{aligned}
&u^2(P') = \left(-\tfrac{1}{T}P'\right)^2 u^2(T) + \left(-\tfrac{1}{f_s}P'\right)^2 u^2(f_s) + \\
&+\left(\tfrac{V_{range}}{2^{b+1}}\tfrac{1}{\sqrt{3}}\right)^2\left(\tfrac{2}{T}\tfrac{1}{f_s}\right)^2\sum_{i=0}^{N-1} I_i^2 + \left(\tfrac{I_{range}}{2^{b+1}}\tfrac{1}{\sqrt{3}}\right)^2\left(\tfrac{2}{T}\tfrac{1}{f_s}\right)^2\sum_{i=0}^{N-1} V_i^2
\end{aligned}
\tag{1.16}
$$

Considering the existence of the system error (1.12), the standard measurement uncertainty of the active power P can be determined using the following expression:

$$u^2(P) = u^2(P') + \left\{ \frac{\partial}{\partial T}\left[\frac{1}{12}\left(\frac{1}{f_s}\right)^2 \frac{2}{T}\int_0^{\frac{T}{2}} \frac{d^2}{dt^2} p(t)\, dt \right] \right\}^2 u^2(T) +$$
$$+ \left\{ \frac{\partial}{\partial f_s}\left[\frac{1}{12}\left(\frac{1}{f_s}\right)^2 \frac{2}{T}\int_0^{\frac{T}{2}} \frac{d^2}{dt^2} p(t)\, dt \right] \right\}^2 u^2(f_s) = \tag{1.17}$$
$$= u^2(P') + \left[\frac{p'(0)-p'(\frac{T}{2})}{6f_s^2T^2} + \frac{p''(\frac{T}{2})}{12f_sT} \right]^2 u^2(T) + \left[\frac{p'(0)-p'(\frac{T}{2})}{3f_s^3T} \right]^2 u^2(f_s)$$

When processing a signal from a power-supply network operating at a specific frequency, the standard measurement uncertainty $u(T)$ is derived from the frequency standard of the power supply. Assuming we have $N + 2$ consecutive voltage samples $v(t)$, encompassing two consecutive zero-crossings, the period T can be evaluated as follows:

$$T = \frac{2}{f_s}\left[(N-1) + \frac{V_0}{V_0 - V_{N-1}} + \frac{V_{N-1}}{V_{N-1} - V_N} \right] \tag{1.18}$$

The measurement uncertainty of T [23] is:

$$u^2(T) = \left(\frac{\partial T}{\partial f_s}\right)^2 u^2(f_s) + \left(\frac{\partial T}{\partial V_{-1}}\right)^2 u^2(V_{-1}) + \left(\frac{\partial T}{\partial V_0}\right)^2 u^2(V_0) +$$
$$\left(\frac{\partial T}{\partial V_{N-1}}\right)^2 u^2(V_{N-1}) + \left(\frac{\partial T}{\partial V_N}\right)^2 u^2(V_N) = \left(\frac{-1}{f_s} T\right)^2 u^2(f_s) + \tag{1.19}$$
$$\left\{ \left[\frac{2}{f_s}\frac{V_0}{(V_0 - V_{-1})^2} \right]^2 + \left[\frac{2}{f_s}\left(\frac{1}{V_0 - V_{-1}} - \frac{V_0}{(V_0 - V_{-1})^2} \right) \right]^2 \right\} 2u^2(V)$$

where $u(V)$ is measuring uncertainties of voltage samples V_i ($V_a \approx V_{range}$).

The relative uncertainty in measuring the period T can be obtained using the following expression:

$$\left(\frac{u(T)}{T}\right)^2 \approx \left(\frac{u(f_s)}{f_s}\right)^2 + \left(\frac{\sqrt{2}}{\pi}\right)^2 \left(\frac{1}{2^b}\frac{1}{\sqrt{3}}\right)^2 \tag{1.20}$$

(b is the number of "accurate" bits).

The standard measurement uncertainty $u(f_s)$ is determined based on the error bounds associated with the frequency of the sampling signal. In the case of a frequency within the range of 1 kHz, this error typically remains below ±0.1 Hz, and in many cases, it can be even lower. As a result, the contributions to the overall measurement uncertainty arising from the uncertainties in T and f_s are significantly smaller compared to the uncertainties attributed to the voltage and current signal samples. It is worth noting that both the active power and the measurement uncertainty are influenced by the function $p(t)$.

1.3 Simulation and Experimental Verification of the Proposed Algorithm

Additional validation of the proposed algorithm was performed through simulation using the Matlab and SIMULINK software package (version R2010b), as depicted in Fig. 1.2. While SPICE simulations offer high precision, they are computationally intensive, particularly for narrow-band modulators with high resolution, due to lengthy period cycles and demanding accuracy requirements. Consequently, selecting an optimized architecture and estimating the building block specifications becomes a time-consuming task in transistor-level design and SPICE simulation. Therefore, there is a need for a simulation environment that is both time-efficient and accurate. With this in mind, the versatile and user-friendly Simulink tool was chosen to develop detailed models of the modulator's building blocks. Simulink proved to be an excellent time-efficient candidate for this purpose.

Since measurements are susceptible to noise, the processing task becomes an estimation problem, meaning that the processed signals may vary depending on the specific noise realization. We address the impact of noise and jitter on the proposed processing method. The presence of noise and jitter can lead to false detection of voltage zero-crossing moments, resulting in incorrect calculations of active power (erroneous determination of integral boundaries). Additionally, errors arise in determining the values of the processed signal samples. These errors originate from imprecise time sampling and the inherent limitations of the real ADC.

As the proposed algorithm is primarily based on standard hardware components employed for signal sampling, the simulation model used in this study reflects the system depicted in Fig. 1.2. A sigma-delta ADC, along with its corresponding sample-and-hold circuit, was utilized for signal conversion [34]. Among oversampling converters, sigma-delta converters have gained significant attention in high-resolution applications due to their noise-shaping characteristics, which result in superior linearity, straightforward implementation, and reduced sensitivity to circuit imperfections. Such converters alleviate the need for complex analog circuitry and, owing to their oversampling nature, are particularly suitable for accurate low- to moderately high-frequency applications.

Sigma-delta modulators can be realized using either the continuous-time (CT) or switched-capacitor (SC) approach. While CT modulators offer advantages such as lower power consumption, higher speed, and intrinsic anti-aliasing filtering, they pose challenges in terms of design complexity, sensitivity to clock jitter, and excessive loop delay. Conversely, SC modulators are preferred over CT modulators for implementation due to their more efficient realization in standard CMOS technology. Moreover, SC modulators enable highly controllable designs and exhibit enhanced robustness against clock jitter and feedback delay issues. For these reasons, the SC sigma-delta modulator is considered in this monography.

The accuracy of calculating active power and RMS value heavily relies on the precision in determining the samples of the voltage and current signals. Hence, the

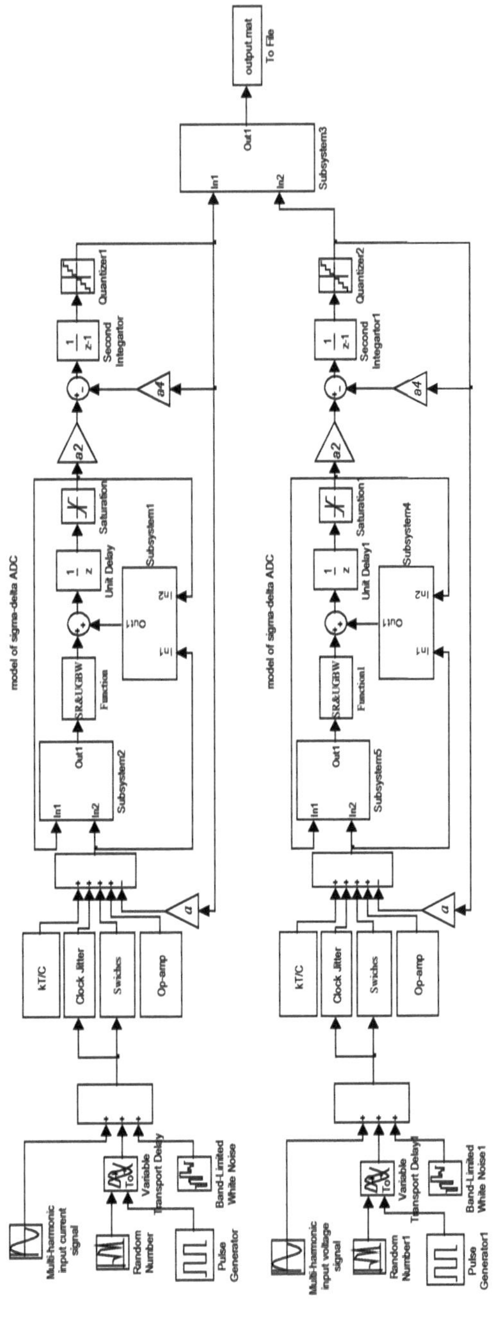

Fig. 1.2 SIMULINK model of the circuit for the realization of proposed algorithm

ADC is simulated with careful consideration of all non-ideal factors associated with the conversion process.

To simulate real-world scenarios, the input signals are generated as a superposition of harmonic components within the "Multi-harmonic input signal blocks," as illustrated in Fig. 1.2. Noise signals are then added to the processed voltage and current signals, simulating possible fluctuations and transient disturbances, thus enhancing the realism of the simulation. As a result, the voltage and current signals exhibit random variations around their nominal values.

The presence of noise and jitter can lead to false detection of signal zero-crossing moments. To account for this, a separate model is employed to simulate the impact of noise and jitter on the voltage and current signals processed by the proposed algorithm. It is important to note that all Taylor expansions of jitter are biased [35], necessitating the inclusion of variable delays in the signal path. This is achieved using a model consisting of a Pulse Generator, Random Number generator, and Variable Transport Delay circuit. Figure 1.2 illustrates the addition of white Gaussian noise to the complex periodic signal. In the simulation model, all possible noise sources (primarily the contributions of operational amplifiers and voltage references) are assumed to be white. The noise power in the "Band-Limited White Noise" block is determined by the Power Spectral Density (PSD), expressed in V^2/Hz.

By employing the Relay block shown in Fig. 1.2, the voltage signal, formed as described above, is transformed into a series of impulses. This enables the detection of zero crossings and the estimation of signal frequency within the "S-Function" block.

1.3.1 Model of Sigma-Delta ADC

To ensure a highly accurate simulation of the conversion circuit, a sigma-delta ADC model described in [34] was utilized. The simulation considered an effective ADC resolution of 16 bits and a sampling rate of $f_S = 100.7$ kHz.

To provide a comprehensive understanding of the simulation model's performance, Fig. 1.3 presents detailed models of clock jitter noise, *kT/C* noise, and the op-amp's thermal and flicker noise at low frequencies. Considering that noise power is additive, the Power Spectral Density (PSD) can be regarded as the sum of a term attributed to flicker (*1/f*) noise and another associated with thermal noise. The thermal noise originates from the sampling switches and the intrinsic noise of the operational amplifier. In this context, k, T, and C represent Boltzmann's constant, temperature in Kelvin, and the sampling capacitor, respectively. V_n denotes the input-referred thermal noise of the op-amp, which comprises contributions from flicker noise, wide-band thermal noise, and DC offset. The total noise power, ${V^2}_n$, can be evaluated through transistor-level simulation. For the purposes of the simulation, V_n was assumed to be 30 μVrms, while the sampling capacitance was set to $C = 2$ pF.

The effect of clock jitter was simulated separately using a model depicted in Fig. 1.3. It was assumed that the time jitter follows an uncorrelated Gaussian random

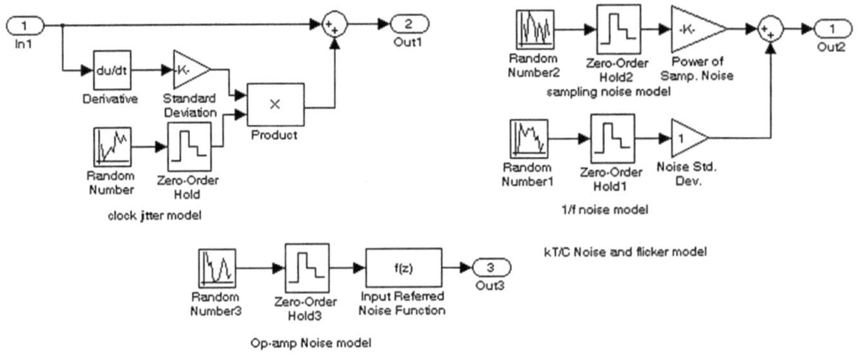

Fig. 1.3 SIMULINK models of clock jitter, kT/C and flicker noise, and op-amp noise

process with a standard deviation $\Delta\tau$. The model of jitter was extensively described and analyzed in [28], demonstrating its dependence on the input sinusoidal frequency. The total in-band error power decreases either by increasing the oversampling ratio (OSR) or by reducing the input bandwidth. In the simulation, the value of jitter was set to 1 ns (standard deviation $\Delta\tau$). Throughout the simulation conducted in this manner, the output PSD of the ideal, thermal noise-affected, and clock jitter-affected signals ranged from -100 to -170 dB for a signal-to-noise distortion ratio (SNDR) between 85 and 96 dB.

The "Integrator DC Gain" block within the model represents the finite DC-Gain (DCG) of the integrator. The presence of finite DCG shifts the pole of the ideal integrator from DC to another frequency and alters the integrator's gain. This phenomenon, known as leakage in the integrator, is depicted by a separate block in the developed model. The nonlinearity of the integrator's DCG, which depends on the output voltage (as shown in Fig. 1.3), introduces distortion. The settling behavior of operational amplifiers (op-amps) comprises two distinct components: the slew rate (SR) and the unity-gain bandwidth (UGBW). These aspects are modeled separately since they affect the settling behavior of the integrator linearly or nonlinearly [26]. The coefficients of the amplifier circuits in Fig. 1.2 are assigned the values $a = 1$; $a_2 = 0.5$; $a_4 = 0.25$.

Switches play a vital role in switched-capacitor (SC) circuits like the sigma-delta ADC. In CMOS technology, switches are implemented using nMOS and pMOS transistors, which exhibit non-ideal characteristics such as nonlinear on-resistance, clock-feed through, and charge injection, the latter of which decreases the sample-and-hold (S/H) time constant [26]. The "Switches charge-injection" block encompasses all of these non-idealities.

To incorporate quantization effects, a quantizer circuit is included in the ADC model using a dedicated block. Quantization error poses a significant challenge in simulations of this nature, especially when using a sigma-delta ADC that shapes the frequency spectrum of the quantization noise. After quantization and sampling, the output signal takes the form of a bit-stream, comprising two components: the low-pass filtered analog input signal and the high-pass filtered quantization noise. The power of the noise is primarily concentrated in the high-frequency region, resulting in a reshaping of the noise spectrum. In the digital domain, the signal undergoes further low-pass filtering and decimation, which effectively cancels out a substantial portion of the quantization noise along with the high-frequency region. Passing through a strong digital low-pass filter, the quantized signal is represented by a greater number of levels than before.

1.3.2 Simulation Results

After determining the values of the input instantaneous power signal samples using the aforementioned ADC model, these values were incorporated into the output file (Fig. 1.2) along with information about the measured frequency of the input voltage signal. This setup enables the calculation of power using the suggested processing algorithms. The MATLAB/Simulink model and code developed can be easily implemented on the dSPACE 1104 platform.

Throughout the simulation, the value of jitter was set to 1 ns (standard deviation $\Delta\tau$). Using this configuration, the output Power Spectral Density (PSD) of the ideal, thermal noise-affected, and clock jitter-affected signals ranged from -80 to -150 dB for a signal-to-noise distortion ratio (SNDR) between 55 and 76 dB. To account for quantization effects, a quantizer circuit was introduced in the ADC model using a dedicated block. Quantization error is a significant concern in simulations of this nature, particularly when employing a sigma-delta ADC that shapes the frequency spectrum of quantization noise.

Upon determining the input signal samples using the aforementioned ADC model, they were fed into the "Subsystem3" block (Fig. 1.2) for the calculation of active power using the suggested processing algorithms. The simulation was performed using the parameters listed in Table 1.1. Given the chosen option of an ADC with a sampling frequency of $f_S = 100.7$ kHz, the processing was conducted with 1012 samples per semi-period (with slight variations depending on detected zero-crossings). As per the derived relation (1.11), the resulting error in the calculation of active power based on the proposed algorithm is inversely proportional to the square of the number of samples, while the error boundaries are specified in Table 1.1. The approximated integral calculations obtained using the trapezium method are resilient to changes in signal frequency due to the sufficiently high sampling frequency.

Table 1.1 follows the notation conventions established in Eq. (1.1), where values in comma-separated columns correspond to harmonics of different orders. For example, in the column for l_s values, 1, 0.34, 0.22, 0.17 correspond to $l_1 = 1$; $l_3 = 0.34$; $l_5 = 0.22$; $l_7 = 0.17$, and so on. Similarly, the amplitude of the third harmonic of the current signal is determined as the product of parameters $l_3 = 0.34$ and $I_R = 10$ A, and so forth. The notation in Table 1.1 aligns with Eq. (1.1). The active power value P is determined based on the definition formula [24]. This value is utilized to evaluate

Table 1.1 Results obtained through a check of the suggested method of measuring ($V_R = 220$ V; $I_R = 10$ A, $f = 50$ Hz)

Number of measurements	k_r	l_s	ψ_r (rad)	ϕ_s (rad)	Relative error of measured value of active power (%)	Relative error of measured RMS value of voltage (%)	Relative error of measured RMS value of current (%)
1	1; 0.56; 0.23	1; 0.34; 0.22; 0.17	0; π; π/6	π/4; π/7; 0; π/2	0.0238	0.0174	0.0197
2	1	1; 0.76; 0.45; 0.36; 0.18	0	0; π; π/12; π/3; π/11	0.0182	0.0113	0.0168
3	1; 0.78	0.98; 0.67; 0.48; 0.36; 0.23	0; π/6	π/4; 0; 0; π/15; π	0.0146	0.0092	0.0124
4	1; 0.43; 0.34	1; 0.84; 0.53	0; 0; π/7	0; π; π/8	0.0239	0.0184	0.0218
5	1	1; 0.77; 0.63; 0.56; 0.49; 0.32; 0.17	π	π/3; π; π/13; π/4; 0; π/11; π/6	0.0243	0.0175	0.0214

the relative error in the calculation of active power P' using the proposed algorithm, Eq. (1.9).

In this simulation setup, the superimposed noise and jitter result in a relative detection error at the fundamental frequency of 0.001%. The simulations indicate that the method is susceptible to noise in the voltage signal but immune to noise in the current signal. The influence of noise is dependent on the sampling frequency, with higher sampling frequencies exhibiting greater susceptibility to noise due to the possibility of detecting one of the dummy signal zero-crossings introduced by added noise. The conducted simulations assessed the feasibility of processing based on the proposed algorithm and a 12-bit resolution ADC, while increasing the oversampling ratio (OSR), and demonstrated consistent performance and error levels.

Table 1.1 clearly demonstrates the significant advantages offered by the simple solution proposed in this paper compared to the complex and costly hardware described in [27, 28], which yielded calculation errors of no less than ±0.1% for the observed quantities. Moreover, our proposed method outperforms the approach in [36] in terms of accuracy in processing AC values, as the frequency domain calculation in [36] resulted in an error of 0.13%.

1.3.3 Experimental Verification

To validate the simulation results, experimental tests were conducted using a laboratory prototype of a "star" connected induction machine. The control structure consists of six thyristors, with each pair of back-to-back connected thyristors linking the phase source voltage to the corresponding phase of the induction machine in the "star" configuration. The parameters of the induction machine are as follows: wound rotor type, two pole pairs, $R_s = 5.2\,\Omega,\ R_r = 14.63\,\Omega,\ L_s = L_r = 0.055\,H,\ L_m = 1.3\,H$ and a synchronous speed of $n = 1500$ rpm.

The experimental setup involved voltage and current transducers, a connector block, a data acquisition card, and a processing computer (PC). The NI ELVIS/PCI-6251 Bundle acquisition card from National Instruments was utilized to capture the current and voltage signals at each phase, which were then processed using the proposed algorithm on the PC. LEM HTR current transducers and LEM CV3 voltage transducers were employed as measurement hardware. The sampling rate of the data acquisition (DAQ) card was set to 250 kS/s.

For the "star" connection at a speed of 1500 rpm, the current and voltage harmonic parameters in phase 'a' were recorded using NI LabVIEW software, as shown in Table 1.2. Signals with this harmonic content were processed using the proposed algorithm to calculate the active power and RMS values of voltage and current. The

Table 1.2 Measured values of voltage and current signals parameters in experimental setup

Parameters	$\alpha = 110°$		$\alpha = 115°$	
	Measured values	Phase (deg)	Measured values	Phase (deg)
I_{ph_a1} (1st harmonic) (A)	0.63	57.87	0.58	32.16
I_{ph_a3} (3rd harmonic) (A)	0	0	0.02	2.15
I_{ph_a5} (5th harmonic) (A)	0.13	−21.77	0.22	18.64
I_{ph_a7} (7th harmonic) (A)	0.04	−80.21	0.18	−128.72
I_{ph_a11} (11th harmonic) (A)	0.1	154.85	0.08	−34.27
I_{ph_a13} (13th harmonic) (A)	0.114	25.47	0.13	143.83
I_{ph_a17} (17th harmonic) (A)	0.02	−12.63	0	0
I_{ph_a19} (19th harmonic) (A)	0.029	−56.48	0.03	−57.47
V_{out_a1} (1th harmonic) (V)	271.94	0	274.32	0
V_{out_a3} (3rd harmonic) (V)	33.62	122.23	28.35	−23.67
V_{out_a5} (5th harmonic) (V)	23.12	−42.37	17.86	63.21
V_{out_a7} (7th harmonic) (V)	10.52	74.68	7.32	−137.42

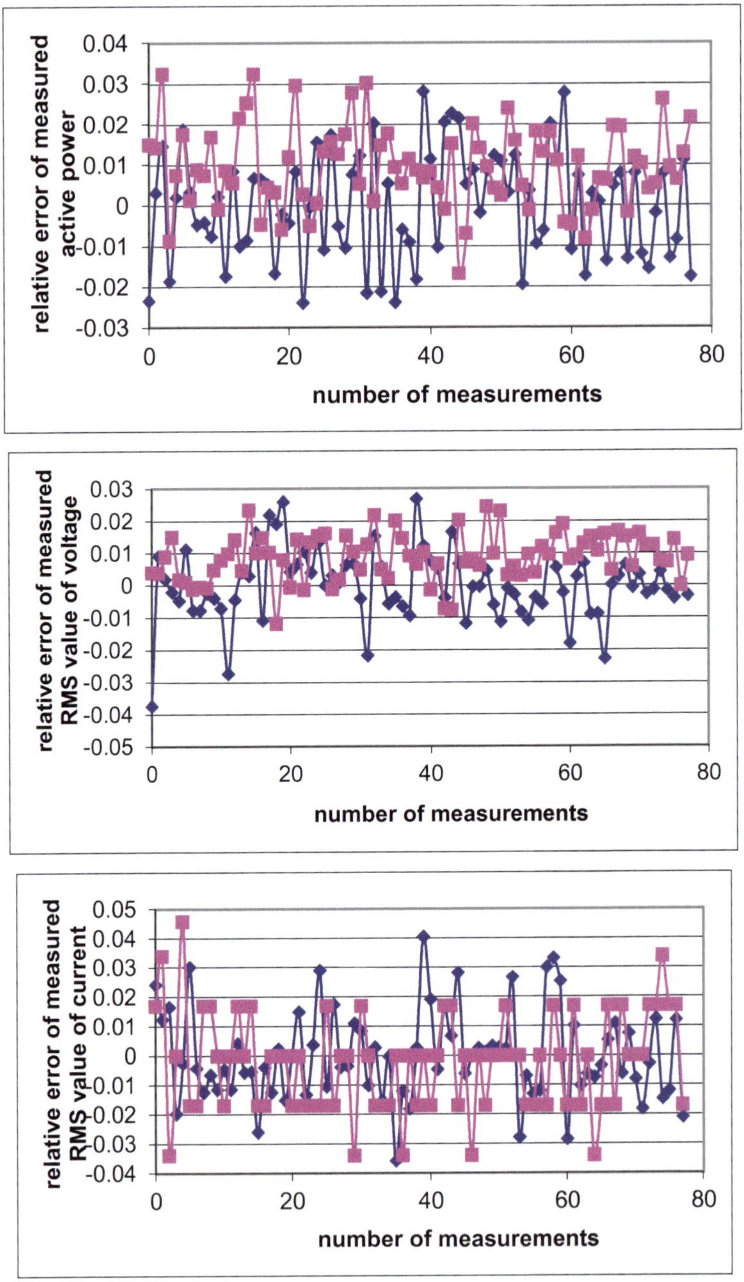

Fig. 1.4 Experimental results (—■—for α = 110°; —◆—for α = 115°)

results of these measurements, depicted in Fig. 1.2, exhibit significantly better performance compared to methods based on frequency domain measurements [37], which exhibited a measurement error in the range of 0.2%. Concurrently, parallel measurements of RMS voltage and current were conducted using a highly precise Fluke 8845 A multimeter to explore the feasibility of employing the proposed algorithm in the development of a new multimeter [38]. Discrepancies between the simulation and experimental results can be attributed to uncertainties in the measurement process, A/D converter errors, and transducer inaccuracies.

Based on the obtained results (Fig. 1.4), it can be inferred that the described procedure serves as a sufficiently accurate method for determining the active power and RMS values of complex AC signals. The proposed measuring algorithm is suitable for implementation in general-purpose microprocessors. In contrast to the algorithm presented in [39], which relies on a complex modification of the Discrete Fourier Transform (DFT) algorithm, the proposed algorithm exhibits a simple structure and overcomes the limitations associated with frequency domain processing.

Through time-domain analysis of the total instantaneous power $p(t)$, it is evident that the active power in public distribution networks can be calculated using Eq. (1.9) at the semi-period of the input voltage signal, assuming the processed signals contain odd harmonic components. The new method requires fewer computations compared to existing methods, which is a significant advantage when considering its implementation and comparing the results to those presented in [7, 27, 28, 39]. Moreover, the proposed algorithm does not require synchronization. The analysis demonstrates that the proposed algorithm maintains a high level of accuracy when processing periodic signals in real-world environments. The precision limit of the algorithm was investigated through theoretical analysis, experimental testing, and simulation. Experimental findings revealed a real-world precision limit of 0.05% in laboratory conditions.

References

1. Bucci, G., Landi, C.: On-line digital measurement for the quality analysis of power systems under nonsinusoidal conditions. IEEE Trans. Instrum. Meas. **48**(2), 853–857 (1999)
2. Carbone, P., Petri, D.: Average power estimation under nonsinusoidal conditions. IEEE Trans. Instrum. Meas. **49**(2), 333–336 (2000)
3. Routray, A., Pradhan, K., Rao, K.P.: A novel Kalman filter for frequency estimation of distorted signals in power systems. IEEE Trans. Instrum. Meas. **51**(3), 469–479 (2002)
4. Jiekang, W., Jun, L., Jixiang, W.: High-accuracy, wide-range frequency estimation methods for power system signals under nonsinusoidal conditions. IEEE Trans. Power Deliv. **20**(1), 366–374 (2005)
5. Bollen, M.H.G., Gu, I.Y.-H.: Signal Processing of Power Quality Disturbances. Wiley-IEEE (2006)
6. Croes, T., Gherasim, C., den Keybus, J.V., Ghijselen, J., Driesen, J., Belmans, R.: Power measurement using the wavelet transform of analytic signals. In: Proceedings of 11th International Conference on Harmonics and Quality of Power, pp. 338–342 (2004)
7. Hamid, E.Y., Mardiana, R., Kawasaki, Z.I.: Method for RMS and power measurements based on the wavelet packet transform. IEE Proc.-Sci., Meas. Technol. **149**(2), 1350–2344 (2002)

8. Clarke, J.J., Stockton, J.R.: Principles and theory of wattmeters operating on the basis of regularly spaced sample pairs. J. Phys. E: Sci. Instrum. **15**, 645–652 (1982)
9. Petrovic, P., Stevanovic, M.: Digital Processing and Reconstruction of Complex AC Signals. Springer (2009)
10. Arpaia, P., Avallone, F., Baccigalupi, A., Capua, C.D.: Real-time algorithms for active power measurements on PWM-based electric drives. IEEE Trans. Instrum. Meas. **45**(2), 462–466 (1996)
11. Petrovic, P.: New method for processing of basic electric values. Meas., Sci. Technol. **19**, 115103 (2008). https://doi.org/10.1088/0957-0233/19/11/115103
12. Pogliano, U.: Use of integrative analog-to-digital converters for high-precision measurement of electric power. IEEE Trans. Instrum. Meas. **50**(5), 1315–1318 (2001)
13. Ramm, G., Moser, H., Braun, A.: A new scheme for generating and measuring active, reactive, and apparent power at power frequencies with uncertainties of 2.5×10^{-6}. IEEE Trans. Instrum. Meas. **48**(2), 422–426 (1999)
14. Germer, H.: High-precision AC measurements using the Monte Carlo method. IEEE Trans. Instrum. Meas. **50**, 457–460 (2001)
15. Langella, R., Testa, A.: The effects of integration intervals on recursive RMS value and power measurements in nonsinusoidal conditions. IEEE Trans. Instrum. Meas. **60**(9), 3047–3057 (2011)
16. Vizireanu, D.N., Halunga, S.V.: Analytical formula for three points sinusoidal signals amplitude estimation errors. Int. J. Electron. **99**(1), 149–151 (2012)
17. Vizireanu, D.N., Halunga, S.V.: Single sine wave parameters estimation method based on four equally spaced samples. Int. J. Electron. **98**(7), 941–948 (2011)
18. Vizireanu, D.N.: A simple and precise real-time four point single sinusoid signals instantaneous frequency estimation method for portable DSP based instrumentation. Measurement **44**(2), 500–502 (2011)
19. Vizireanu, D.N.: Quantized sine signals estimation algorithm for portable DSP based instrumentation. Int. J. Electron. **96**(11), 1175–1181 (2009)
20. Udrea, R.M., Vizireanu, D.N.: Quantized multiple sinusoids signal estimation algorithm. J. Instrum., JINST **3**, PO2008 (2008)
21. Petrović, P., Župunski, I.: A new time-domain method for calculation of basic electric values in public distribution networks. Trans. Inst. Meas. Control **35**(3), 279–288 (2013)
22. Petrovic, P.: Modified formula for calculation of active power and RMS value of band-limited AC signals. IET Sci. Meas. Technol. **6**(6), 510–518 (2012)
23. IEC 50160: Voltage characteristics of electricity supplied by public distribution networks (2011)
24. IEEE Std. 1459-2000: Definitions for the measurement of electric quantities under sinusoidal, non-sinusoidal, balanced or unbalanced conditions (2000)
25. Yazici, I., Ozdemir, A.: Hardware-based simple, fast and accurate measurement of power system frequency using a hybrid system. Trans. Inst. Meas. Control **33**(8), 1004–1022 (2011)
26. Kunjumuhammed, L.P., Mishra, M.K.: A control algorithm for single-phase active power filter under non-stiff voltage source. IEEE Trans. Power Electron. **21**(3), 822–825 (2006)
27. Martens, O., Trampark, H., Liimets, A., Nobel, P., Veskimester, A., Jarvalt, A.: DSP-based power quality monitoring device. In: IEEE International Symposium on Intelligent Signal Processing, pp. 1–5 (2008)
28. Hindersah, H., Purwadi, A., Ali, F.Y., Heryana, N.: Prototype development of single phase prepaid KWh meter. In: International Conference on Electrical Engineering and Informatics, pp. 1–6 (2011)
29. Kuseljević, M.D.: A simultaneous estimation of frequency, magnitude, and active and reactive power by using decoupled modules. IEEE Trans. Instrum. Meas. **59**(7), 1866–1873 (2010)
30. Numerical Integration: Curse Materials. Dept. Math. Sci.—Worcester Polytech. Inst., Worcester, MA. http://www.math.wpi.edu/Course_Materials/MA1023C05/num_int/node1.html (2005)
31. Novotny, M., Sedlacek, M.: RMS value measurement based on classical and modified digital signal processing algorithms. Measurement **41**, 236–250 (2008)

32. Pogliano, U., Trinchera, B., Francone, F.: Reconfigurable unit for precise RMS measurements. IEEE Trans. Instrum. Meas. **58**(4), 827–831 (2009)
33. ISO 1993: Guide to the expression of uncertainty in measurement. ISO, Geneva
34. Hoseini, H.Z., Kale, I., Shoaei, O.: Modeling of switched-capacitor delta-sigma modulators in SIMULINK. IEEE Trans. Instrum. Meas. **54**(4), 1646–1654 (2005)
35. Coakley, K.J., Wang, C.M., Hale, P.D., Clement, T.S.: Adaptive characterization of jitter noise in sampled high-speed signals. IEEE Trans. Instrum. Meas. **52**(5), 1537–1547 (2003)
36. Asquerino, J.C.M., Ibanez, M.C., Ojeda, A.L., Benitez, J.G.: Measurement of apparent power components in the frequency domain. IEEE Trans. Instrum. Meas. **39**(4), 583–587 (1990)
37. Cristaldi, L., Ferrero, A., Ottoboni, R.: Measuring equipment for the electric quantities at the terminals of an inverter-fed induction motor. IEEE Trans. Instrum. Meas. **45**(2), 449–452 (1996)
38. http://us.fluke.com/usen/products/
39. Odzemir, A., Ferikoglu, A.: Low cost mixed-signal microcontroller based power measurement technique. IEE Proc.-Sci., Meas. Technol. **151**(4), 253–258 (2004)

Chapter 2
A Simple Algorithm for Simultaneous Sine Signal Parameters Estimation

The scientific literature addressing electrical parameter measurement techniques for power system applications is extensive and readily accessible. Accurately estimating parameters such as frequencies, magnitudes, and phases of periodic signals embedded in noise is crucial in various practical applications, including signal processing, system identification, and control system design. Magnitude estimation of power system signals has been a significant area of research for several decades, with established methods for signals of known frequencies. Measuring electrical parameters of a fixed-frequency signal is a straightforward task [1, 2]. However, when the frequency is unknown in advance, achieving precise measurements of amplitude and phase becomes highly challenging. Numerical algorithms employed for power measurements are often sensitive to frequency variations, such as those assumed in fast Fourier transform (FFT) or least-mean-square (LMS) techniques, where the system frequency is assumed to be constant and predetermined (e.g., 50 or 60 Hz) [3, 4].

In an electric power system, deviations in frequency occur due to disturbances such as large-scale load connections or disconnections and offline generation sources. Frequency variations are more likely to occur in loads supplied by generators isolated from the utility systems (islands). Any deviation from the nominal frequency of 50 or 60 Hz can significantly impact the performance of measurement devices that assume a constant frequency.

Spectrum estimation of discretely sampled processes typically relies on procedures employing the fast Fourier transform (FFT), an efficient computational algorithm for computing the discrete Fourier transformation (DFT). However, while the FFT is effective under fixed-frequency conditions, its performance diminishes unless the sampling frequency and the fundamental frequency of the signal are synchronized. It is well-known that the accuracy of the FFT is compromised under desynchronization and nonstationary conditions, where the fundamental or harmonic frequencies may vary over time. This discrepancy arises from the finite-impulse-response (FIR) filters in the FFT process, which exhibit different magnitude gains at frequencies other than the nominal power frequency [3]. Additionally, the frequencies of

P. Petrović, *Processing, Estimation and Measurement of Signals Parameters in Public Distribution Networks*, Studies in Systems, Decision and Control 502,
https://doi.org/10.1007/978-3-031-43107-4_2

harmonics align with the zero points of the frequency response of the FIR filter when a rectangular window is used in DFT algorithms. These limitations become particularly problematic when analyzing short data records, a common occurrence in practical scenarios due to the brevity of many measured processes.

To enhance the periodicity requirement of the FFT process, time-weighting functions known as windows or correction interpolation algorithms are employed [4]. While these methods reduce errors, they do not eliminate them entirely. If no window is used, synchronization with the grid's fundamental frequency becomes essential. Unfortunately, traditional synchronization techniques like the phase-locked loop (PLL) have relatively long response times, especially when transient phenomena such as power supply frequency variations or phase jumps are present.

In addition to the challenges associated with synchronizing the sampling frequency and signal frequency, the fast Fourier transform (FFT) suffers from limitations due to frame implementation. The FFT processes data in entire frames and cannot provide intermediate data points. When applied in a sliding mode, where the FFT is repeatedly computed for a frame of N elements by considering the last $N - 1$ shifted elements from the previous frame along with a single new element, intensive computational effort is required, making integration into low-cost microcontrollers complex.

To address the sensitivity to frequency variations, a Newton-type recursive numerical algorithm has been proposed [5]. This algorithm considers the system frequency as an unknown signal model parameter to be estimated and simultaneously estimates the frequency and spectra of the power system. By incorporating the power frequency into the vector of unknown model parameters, the signal model becomes nonlinear, necessitating the use of nonlinear estimation strategies. The recursive algorithm is further enhanced by employing a strategy of sequential tuning of the forgetting factor, leading to significant improvements in convergence and accuracy.

When the generator and the acquisition device are not synchronized, the least-squares (LS) technique can be utilized to design FIR filters with optimized frequency responses that do not require synchronization [6]. However, this approach results in a higher computational load compared to the synchronized case. The LS design method for large-order filters entails substantial computation, which may exceed the available time within a single sampling interval. Consequently, these filters cannot be efficiently adapted online during frequency deviations. To mitigate the computational burden, an alternative approach involves tabulating the weights appropriately.

A novel technique for simultaneous estimation of the local system frequency, amplitude, and phase over a wide frequency variation range is proposed [7]. The proposed algorithm only requires the sample value of the processed signal along with its first and second derivative values. The framework includes an oversampling analog-to-digital conversion unit with a dithering process, a higher-order FIR digital differentiator, and a decimator. As a result, the algorithm is straightforward and demands modest resources for implementation.

In contrast to the analyzed IEEE standard [8], the proposed algorithm offers significantly enhanced stability and is free from propagation errors. When following the procedure prescribed by the standard, amplitude errors of the fundamental propagate through the method since the amplitudes are used for reconstructing the detected sine wave and determining the results before being employed to determine the parameters of subsequent harmonics. Consequently, frequency and amplitude errors from the initial calculation contaminate the higher harmonics, leading to inevitable inaccuracies in phase and amplitude estimation at each step.

2.1 Suggested Method of Processing

Consider an assumption where the measured signal, be it voltage or current, undergoes a filtering process to minimize and eliminate the influence of noise and higher harmonic components. In this context, the observation model outlined below should be employed:

$$x(t) = X\sin(\omega_i t + \phi_i) \tag{2.1}$$

In the given context, let X represent the magnitude of the processed signal. Upon differentiation of the signal (2.1), we obtain the following expression:

$$\begin{aligned} &\frac{d(x(t))}{dt}\Big|_{t=t_n} = \frac{d(X\sin(\omega_i t + \phi_i))}{dt}\Big|_{t=t_n} = y_1(t_n) = y_1[n] \Rightarrow \\ &y_1(t_n) = \omega_i X\cos(\omega_i t_n + \phi_i) \end{aligned} \tag{2.2}$$

$$\begin{aligned} &\frac{d^2(x(t))}{dt^2}\Big|_{t=t_n} = \frac{d^2(X\sin(\omega_i t + \phi_i))}{dt^2}\Big|_{t=t_n} = y_2(t_n) = y_2[n] \Rightarrow \\ &y_2(t_n) = -\omega_i^2 X\sin(\omega_i t_n + \phi_i) \end{aligned} \tag{2.3}$$

In this context, t_n denotes the arbitrary time moment at which the differentiation of the input analog signal is performed. The differential values-samples obtained from this process can be utilized to calculate the unknown signal parameters using the following expression:

$$\begin{aligned} f_i &= \frac{1}{2\pi}\sqrt{\left|\frac{y_2(t_n)}{x(t_i)}\right|} \\ \phi_i &= arctg\left(\frac{x(t_n)}{y_1(t_n)}2\pi f_i\right) - 2\pi f_i t_n \\ X &= \frac{x(t_n)}{\sin(2\pi f_i t_n + \phi_i)} \end{aligned} \tag{2.4}$$

Figure 2.1 depicts a proposed system configuration for obtaining the first and second-order derivatives in an oversampling setup. The signal of interest is captured by a sensor and then subjected to conditioning through a signal conditioning circuit (amplifier) and band-limited by an anti-aliasing filter. Subsequently, the conditioned analog signal $x(t)$ is combined with dithering noise, enabling the composite signal to be fed into an analog-to-digital converter (ADC) unit operating at an oversampling rate of $f_{sL} = Lf_s$ Hz (samples/s). Here, f_s denotes the minimum sampling rate (Nyquist sampling rate), while L represents the oversampling factor. Each digital sample $x[n]$ is encoded using N_q bits. The first and second-order derivatives of the digitized signal are obtained using first and second-order finite impulse response (FIR) digital differentiators operating at the oversampling rate [5]. These differentiators have transfer functions designed as $H_1(z)$ and $H_2(z)$, respectively. Following decimation by a factor of L, the resulting first and second-order derivative signals $y_1[m]$ and $y_2[m]$ are achieved at the Nyquist rate of f_s Hz. The impact of the digital differentiator on the oversampling rate reshapes the spectrum of quantization noise, effectively pushing it towards the high-frequency range and simultaneously filtering it. Consequently, an improvement in the signal-to-quantization-noise ratio (SQNR) can be expected for the estimated derivative signal after decimation.

The anti-aliasing filter (depicted in Fig. 2.1) has a bandwidth of $f_s/2$ Hz. Although the addition of dithering noise raises the average spectral noise floor of the original input signal, the dithering process introduces incoherence between the quantized error and the original input signal, resulting in a white and flat spectrum for the quantization noise. Thus, the oversampling technique can be effectively employed to compensate for the degraded SQNR and further enhance it by increasing the sampling rate. Typically, random wideband dithering noise, generated by a noise diode or noise generator ICs, possesses a root-mean-square (RMS) level equivalent to 1/3 to 1 least significant bit (LSB) voltage level. The ideal frequency response of the kth-order differentiator $H_k(z)$ is denoted as:

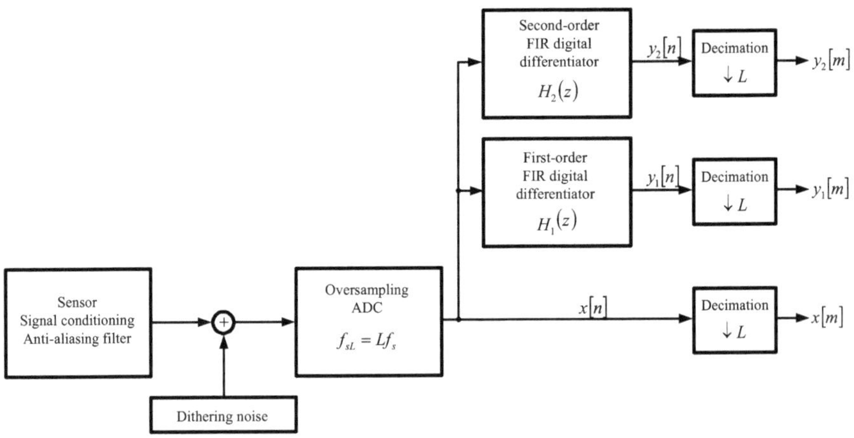

Fig. 2.1 Proposed system for signal reconstruction based on first and second-order differentiators in the over-sampling system

$$H_k\left(e^{j\omega}\right) = \begin{cases} 0, & for \quad -\pi \le \omega \le -\omega_{\max} \\ (j\omega/\omega_{\max})^k & for \quad -\omega_{\max} \le \omega \le \omega_{\max} \\ 0, & for \quad \omega_{\max} \le \omega \le \pi \end{cases} \tag{2.5}$$

In the context of the proposed system, the normalized digital frequency $\omega = 2\pi f/f_{sL}$ in radians is denoted as ω, and the maximum normalized digital frequency of the sensor signal $\omega_{\max} = 2\pi(f_s/2)/f_{sL} = \pi/L$, in radians, is represented by $\omega_{\max}$. Within the oversampling system, $\omega_{\max} << \pi$ and $H_k\left(e^{j\omega}\right)$ is normalized to have a unit gain at $\omega_{\max}$. By employing the ideal frequency domain specification, the kth-order differentiator can be designed and implemented. An effective approach for designing a finite impulse response (FIR) digital differentiator using Fourier transform design is proposed in [9].

2.1.1 Recursion of FIR Differentiator Coefficients

Given the ideal frequency response $H_k\left(e^{j\omega}\right)$, corresponding to the kth-order digital differentiator in Eq. (2.5), we can obtain the noncausal coefficients of the finite impulse response (FIR) differentiator for the kth-order derivative as follows:

$$h_k[n] = \frac{1}{2\pi} \int_{-\omega_{\max}}^{\omega_{\max}} H_k\left(e^{j\omega}\right) e^{jn\omega} d\omega, \; for \; -M \le n \le M \tag{2.6}$$

Here, M represents a positive integer used to truncate the obtained coefficient sequence of the differentiator, resulting in a total of $N = 2M + 1$ coefficients. It should be noted that $\omega_{\max}$ refers to the stop frequency edge. In the case of a full-band differentiator, $\omega_{\max}$ is equal to π radians. By substituting Eq. (2.5) into Eq. (2.6), we can derive the following expression:

$$h_k[n] = \frac{1}{2\pi} \int_{-\omega_{\max}}^{\omega_{\max}} (j\omega/\omega_{\max})^k e^{jn\omega} d\omega, \; for \; -M \le n \le M \tag{2.7}$$

Using integrations by parts, we achieve:

$$h_k[n] = \frac{(j\omega)^k}{jn\omega_{\max}^k} e^{jn\omega} \Big|_{-\omega_{\max}}^{\omega_{\max}} - \frac{k}{n\omega_{\max}^k} \int_{-\omega_{\max}}^{\omega_{\max}} (j\omega)^{k-1} e^{jn\omega} d\omega \tag{2.8}$$

Replaying (2.8) in (2.7), it leads to the general recursion shown as follows:

$$h_k[n] = \frac{(j\omega)^k}{jn\omega_{\max}^k} e^{jn\omega}\Big|_{-\omega_{\max}}^{\omega_{\max}} - \frac{k}{n\omega_{\max}^k} h_{k-1}[n] \tag{2.9}$$

It is possible realize (2.9) in the following recursion. For $n = 0$:

$$h_k[0] = \frac{1}{2\pi} \int_{-\omega_{\max}}^{\omega_{\max}} (j\omega/\omega_{\max})^k d\omega = \begin{cases} \frac{j^k \omega_{\max}}{\pi(k+1)}, & for\ k = evenn \\ 0, & for\ k = odd \end{cases} \tag{2.10}$$

For $n \neq 0$ and $k =$ even (the even-order derivative):

$$h_k[n] = \frac{j^k \sin(n\omega_{\max})}{n\pi} - \frac{k}{n\omega_{\max}} h_{k-1}[n] \tag{2.11}$$

For $n \neq 0$ and $k =$ odd (the odd-order derivative):

$$h_k[n] = \frac{j^{(k-1)} \cos(n\omega_{\max})}{n\pi} - \frac{k}{n\omega_{\max}} h_{k-1}[n] \tag{2.12}$$

Note that the recursion starts from a low-pass filter given by:

$$h_0[n] = \frac{\sin(n\omega_{\max})}{n\pi} \tag{2.13}$$

Hence, starting from Eq. (2.13), we can calculate the noncausal coefficients of the FIR digital differentiators by employing Eq. (2.10) along with (2.11) for the second-order derivative or Eqs. (2.10) and (2.12) for the first-order derivative. To obtain the causal differentiator coefficients, we can shift the noncausal filter coefficients by M samples. Additionally, in order to mitigate the ripple effect [5] resulting from truncation of the filter sequence, a window function is applied to obtain the windowed filter coefficients. The overall procedure for designing the FIR digital differentiator can be summarized as follows:

- Given the length of the FIR digital differentiator with $N = 2M + 1$ (an odd integer) and the kth-order derivative, compute the noncausal filter coefficients $h_k[n]$ using Eqs. (2.10), (2.11) or (2.12), and (2.13) for $n = 0, 1, 2, \ldots, M$.
- Apply the window function to achieve the windowed filter coefficients for $n = 0, 1, 2, \ldots, M$ according to Eq. (2.14).

$$\overline{h}_k[n] = h_k[n]w[n] \tag{2.14}$$

- Use the coefficient symmetric properties to acquire the other half of differentiator coefficients. For $k =$ even:
- Utilize the coefficient symmetric properties to obtain the other half of the differentiator coefficients. For $k =$ even, use Eq. (2.15),

$$\overline{h}_k[n] = \overline{h}_k[-n], \;\; for\ n = -1, -2, \ldots, -M \tag{2.15}$$

- and for $k =$ odd, employ Eq. (2.16).

$$\overline{h}_k[n] = -\overline{h}_k[-n], \;\; for\ n = -1, -2, \ldots, -M \tag{2.16}$$

- Obtain the causal differentiator coefficients by shifting the filter coefficients.

$$h_k[n] = \begin{cases} \overline{h}_k[n-M], & n=0,1,\ldots,M-1 \\ \overline{h}_k[M-n], & n=M,M+1,\ldots,2M+1 \end{cases} \tag{2.17}$$

2.1.2 *SQNR*

The assumption of white and uniformly distributed quantization noise within the quantization interval is often not valid in practice. When quantizing a sinusoidal signal, for example, the quantization noise becomes correlated with the quantized sinusoid, resulting in a degradation of signal-to-quantization-noise ratio (SQNR) performance in the oversampling system. To address this issue, a dithering process is employed, which involves adding white Gaussian noise with an RMS level ranging between 1/3 and 1 least significant bit (LSB) voltage to the analog signal prior to oversampling. This dithering process helps approximate a flat quantization noise spectrum.

In practical scenarios, differentiators typically exhibit a roll-off curve in the transition band, effectively filtering higher frequency components. Consequently, using the ideal frequency response for SQNR analysis may lead to lower SQNR values if there is no requirement for signal components around the cut-off frequency edge.

We assume that the oversampling system employs an N_q-bit analog-to-digital converter (ADC) resolution, and the user can increase the sampling rate in the data acquisition system to achieve the desired oversampling rate. Considering the addition of dithering noise $w_a[n]$ to the analog signal $x_a[n]$, the quantized signal $x[n]$ after ADC can be expressed as:

$$x[n] = Q(x_a[n] + w_a[n]) = x_a[n] + w_a[n] + e[n] \tag{2.18}$$

In the given equation, $Q()$ represents an N_q-bit uniform quantizer, while $e[n]$ denotes the quantization noise. The quantization noise is assumed to be white and uniformly distributed, with its power determined by the following expression:

$$\sigma_q^2 = \frac{(X/2^{N_q-1})^2}{12} \tag{2.19}$$

The quantization error can be defined as the discrepancy between the quantized signal and the analog signal samples without dithering. By calculating the mean-square expectation of this difference, we can evaluate the power of the quantization error as follows:

$$\sigma_{qd}^2 = E\left\{(x[n] - x_a[n])^2\right\} \tag{2.20}$$

Here, $E()$ represents the mean-square expectation operator. Assuming that the dithering noise and quantization noise are uncorrelated, the power of the quantization error with the dithering process can be expressed as the sum of the dithering noise power $\sigma^2{}_d$ and the quantization noise power $\sigma^2{}_q$. After oversampling and differentiation, the power of the in-band quantization error within the frequency range from $-\omega_{\max}$ to $\omega_{\max}$ can be approximately estimated by utilizing the ideal transfer function of the kth-order derivative differentiator:

$$\sigma_{in-band}^2 = \int_{-\omega_{\max}}^{\omega_{\max}} \frac{\sigma_{qd}^2}{2\pi}\left|H_k\left(e^{-j\omega}\right)\right|^2 d\omega = \frac{\omega_{\max}}{\pi(2k+1)}\left(\frac{X^2 2^{-2N_q-2}}{12} + \sigma_d^2\right) \tag{2.21}$$

Substituting $\omega_{\max} = 2\pi f_{\max}/f_{sL} = \pi/L$ in (2.21) the SQNR can be expressed as:

$$SQNR = \frac{12L(2k+1)2^{2N_q-2}/X^2}{\left(1 + 12 \cdot 2^{2N_q-1}\sigma_d^2/X^2\right)}\sigma_k^2 \tag{2.22}$$

In the given equation, $\sigma^2{}_k$ represents the power of the kth-order derivative signal $y_k[m]$. Considering a sinusoidal signal described by Eq. (2.1), we can calculate the power of the kth-order derivative signal as follows:

$$\sigma_k^2 = \frac{1}{2}\left(\frac{X\omega_i^k}{\omega_{\max}^k}\right)^2 = \frac{X^2}{2}\left(\frac{f_i}{f_{\max}}\right)^{2k} \tag{2.23}$$

The introduction of dithering ensures that the spectrum of quantization error remains white and relatively flat across all frequency ranges. At higher sampling rates, the spectrum of in-band signals becomes narrower, causing the quantization error to exhibit a tendency to correlate with the quantized signal. Consequently, the spectrum of the derivative signal contains significant low-frequency noise components that cannot be effectively removed through differential filtering. By employing the dithering process, the spectrum of the quantization error remains white and relatively flat, enabling continuous improvement in signal-to-quantization-noise ratio (SQNR) by increasing the oversampling factor. Furthermore, assuming that the SQNR reaches a plateau at an oversampling factor of L_N without employing the dithering process, we can easily achieve the same SQNR with the dithering process by employing an oversampling factor of $L \geq \left(1 + 12 \cdot 2^{2N_q-1}\sigma_d^2/X^2\right)L_N$. This relationship demonstrates that an increased oversampling factor is necessary to offset the drawbacks associated with the dithering process. In Eq. (2.22), by utilizing $\sigma_d = \Delta$V/3 V (equivalent to 1/3 LSB), where ΔV represents the analog-to-digital converter (ADC) resolution, and considering the signal's full range of $2X = 2^{N_q}\Delta V$ volts, we obtain the relationship $L = (7/3)L_N$.

In practical applications of the proposed algorithm, it is important to consider the presence of errors in determining the samples $x(t_n)$, $y_1(t_n)$, and $y_2(t_n)$. To obtain the best estimate of these values according to the assumed criterion, a recalculation process can be employed iteratively W times (where W is arbitrary). During this process, we form series $x(t_n)_i$, $y_1(t_n)_i$, and $y_2(t_n)_i$ ($i = 1,\ldots,W$), as outlined in the proposed algorithm. The random measurement errors Δ_n are unbiased with $E(\Delta_i) = 0$, have the same variance $\mathrm{var}(\Delta_i) = \sigma^2$, and are uncorrelated with each other. Under these assumptions, the weighted average procedure can be utilized to reduce random errors in determining the observed values. The weighted averages, denoted as $\hat{x}(t_n)$, $\hat{y}_1(t_n)$, $\hat{y}_2(t_n)$, for the values $x(t_n)$, $y_1(t_n)$, and $y_2(t_n)$) are calculated as follows:

$$\hat{x}(t_n) = \frac{\sum_{i=1}^{n_x} w_{xi}x(t_n)_i}{\sum_{i=1}^{n_x} w_{xi}}; \hat{y}_1(t_n) = \frac{\sum_{n=1}^{n_{y_1}} w_{y_1 i}y_1(t_n)}{\sum_{i=1}^{n_{y_1}} w_{y_1 i}}; \hat{y}_2(t_n) = \frac{\sum_{n=1}^{n_{y_2}} w_{y_2 i}y_2(t_n)}{\sum_{i=1}^{n_{y_2}} w_{y_2 i}};$$

$$\sum_{i=1}^{n_x} w_{xi} = \sum_{i=1}^{n_{y_1}} w_{y_1 i} = \sum_{i=1}^{n_{y_2}} w_{y_2 i} = W \tag{2.24}$$

In the provided equation, w_{xi}, $w_{y_1 i}$, $w_{y_2 i}$ are the non-negative weights correspond to the series of $x(t_n)$, $y_1(t_n)$, and $y_2(t_n)$. The parameter determines the number of distinct values in the aforementioned series over W iterations. The selection of the value for W depends on the desired processing speed, where a higher value of W leads to more accurate estimations of the values.

2.2 Simulation Results

The proposed algorithm in this paper has been evaluated through computer simulations using input data under various steady-state, dynamic, and transient operational conditions. The performance of the proposed power component estimation technique has been examined for multiple test cases.

To begin, the algorithm was applied to a sine signal with an 8-bit analog-to-digital converter (ADC) resolution and an oversampling factor L of 256. Dithering noise $\sigma_d = (1/3)LSB$ was added to the analog signal prior to oversampling, resulting in oversampling the sine signal with 4096*L samples. In the first test, an input-frequency modulated sinusoidal signal with a step frequency change from 50 to 49.8 Hz at $t = 0$ s was processed (Fig. 2.2a). White noise with a signal-to-noise ratio (SNR) of 60 dB was added to the sinusoidal signal. The proposed algorithm demonstrated adaptive tracking capabilities for time-varying characteristics of the power signal. As depicted in Fig. 2.2a, the technique provided accurate frequency estimation with an error range of 0.001 Hz, similar to the results obtained in [10].

The algorithm's ability to estimate frequencies over a wide range of frequency changes was investigated using sinusoidal test signals with the time dependence $f(t) = 50 + 0.5\sin(10\pi t)$ as shown in Fig. 2.2b. The dynamic responses observed were significantly better than those presented in [11]. These results further validate the arguments and conclusions from the previous frequency-step-changing case.

Additional testing of the proposed algorithm was performed through simulation using the MATLAB and SIMULINK software packages. The impact of noise on the signal was examined by using a sinusoidal 50 Hz input test signal with additive white Gaussian noise. The random noise was adjusted to achieve a prescribed signal-to-noise ratio (SNR) defined as $SNR = 20\log\left(X/\sqrt{2}\sigma\right)$, where σ represents the noise standard deviation. Figure 2.3 illustrates the maximum errors observed in frequency and harmonic magnitude estimates when input signals of 30, 50, and 70 Hz were used, with SNRs ranging from 40 to 70 dB. It is important to note that in practical scenarios, the SNR of voltage signals obtained from power systems typically falls within the range of 50 to 70 dB. At this noise level, the proposed technique exhibits minimal error, as depicted in Fig. 2.3. The obtained results surpass those presented in [12–14].

The algorithm introduced in [7] presents a novel approach for estimating frequency and signal parameters with reduced complexity. This algorithm allows for simultaneous estimation of the signal parameters while considering the presence of time-varying frequencies. By identifying the parameters of the input analog signals, it becomes possible to determine key values relevant to electric utilities such as energy, power, and RMS values. Simulation results demonstrate that the proposed algorithm achieves satisfactory precision in reconstructing periodic signals within real-world environments.

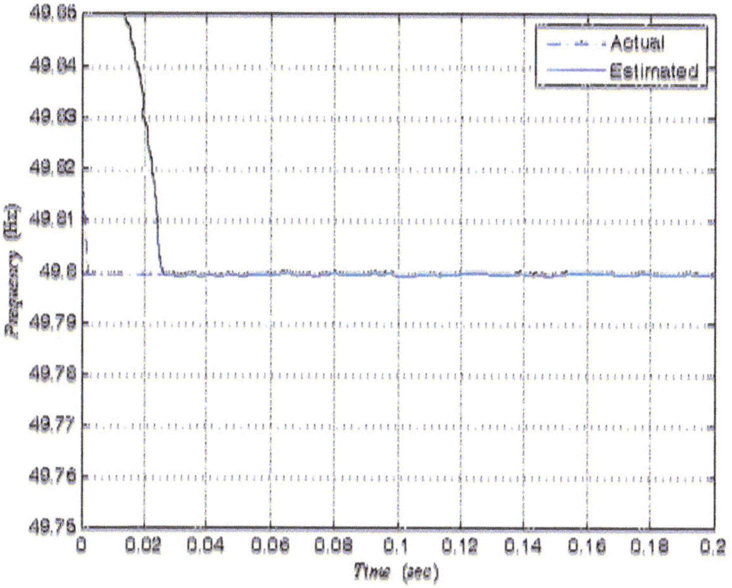

(a)

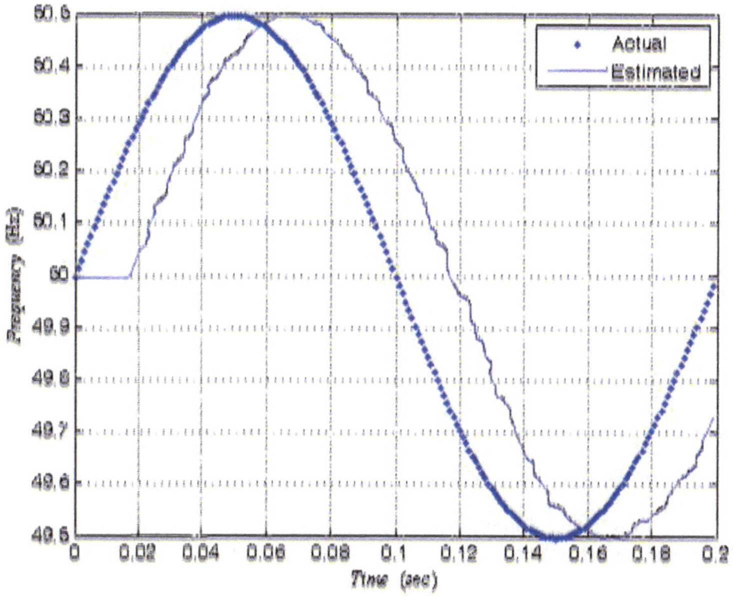

(b)

Fig 2.2 **a** Estimation for $f = 50$ Hz for $t < 0$ s and $f = 49.8$ Hz for $t > 0$ s with SNR = 60 dB and with harmonics presence; **b** estimation for $f(t) = 50 + 0.5\sin(10\pi t)$ with SNR = 60 dB

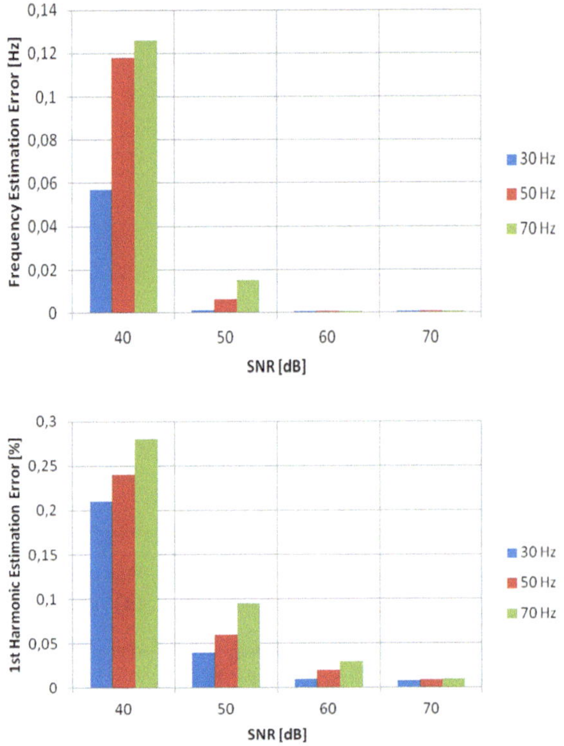

Fig. 2.3 Maximum estimations errors for noisy input signals

References

1. Sekhar, S.C.: Auditory motivated level-crossing approach to instantaneous frequency estimation. IEEE Trans. Signal Process. **53**(4), 1450–1462 (2005)
2. Wang, F., Bollen, M.: Frequency response characteristics and error estimation in RMS measurement. IEEE Trans. Power Deliv. **19**(4), 1569–1578 (2004)
3. Pantazis, Y., Rosec, O., Stylianou, Y.: Iterative estimation of sinusoidal signal parameters. IEEE Signal Process. Lett. **17**(5), 461–464 (2010)
4. Wand, M., Sun, Y.: A practical, precise method for frequency tracking and phasor estimation. IEEE Trans. Power Deliv. **19**(4), 1547–1552 (2004)
5. Terzija, V.V.: Improved recursive Newton-type algorithms for frequency and spectra estimation in power systems. IEEE Trans. Instrum. Meas. **52**(5), 1654–1659 (2003)
6. Sidhu, T.T.: Accurate measurement of power system frequency using a digital signal processing technique. IEEE Trans. Instrum. Meas. **48**(1), 75–81 (1999)
7. Petrovic, P.: A simple algorithm for simultaneous sine signal parameters estimation. J. Electr. Eng. **64**(3), 180–185 (2013)
8. Arpaia, P., Cruz Serra, A., Daponte, P., Monteiro, C.L.: A critical note to IEEE 1057-94 standard on hysteretic ADC dynamic testing. IEEE Trans. Instrum. Meas. **50**(4), 941–948 (2001)
9. Tan, L.: Digital Signal Processing: Fundamentals and Applications. Elsevier, New York (2007)
10. Sarkar, A., Sengupta, S.: Second-degree digital differentiator-based power system frequency estimation under non-sinusoidal conditions. IET Sci. Meas. Technol. **4**(3), 105–114 (2010)
11. Tse, N.C.F., Lai, L.L.: Wavelet-based algorithm for signal analysis. EURASIP J. Adv. Signal Process. **2007**, 38916 (2007)

12. Kuseljevic, M.D.: A simple recursive algorithm for simultaneous magnitude and frequency estimation. IEEE Trans. Instrum. Meas. **57**(7), 1207–1214 (2008)
13. Zheng, J.K., Lui, W.K., Ma, W.-K., So, H.C.: Two simplified recursive Gauss-Newton algorithms for direct amplitude and phase tracking of a real sinusoid. IEEE Signal Process. Lett. **14**(12), 972–975 (2007)
14. Hocanin, A., Kukrer, O.: Estimation of the frequency and waveform of a single-tone sinusoid using an offline-optimized adaptive filter. In: Proceedings of the IEEE International Conference on Acoustics, Speech, and Signal Processing (ICASSP '05), vol. 4, pp. iv/349-iv/352 (2005)

Chapter 3
Algorithm for Fourier Coefficient Estimation

Accurately estimating the amplitude and phase of a signal, even when the frequencies within the signal are known, is of great importance in various fields [1]. In power systems, for instance, accurately estimating the harmonic components is essential for ensuring power supply quality. When the frequencies in the target signal exhibit a harmonic relationship or are integer multiples of a fundamental frequency, the magnitude and phase of the signal can be obtained using well-established methods such as discrete Fourier transform (DFT) and short-time DFT, which are simple and cost-efficient tools for these applications [1]. These techniques allow the extraction of Fourier coefficients at each sampling time using a sliding frequency sampling structure.

Numerous studies have focused on sampling techniques and optimal methods for reconstructing band-limited signals, represented as Fourier series or trigonometric polynomials. One such study [2] introduces a new technique for reconstructing a band-limited signal from periodic nonuniform samples by employing a model based on perfect reconstruction multirate filter bank theory. Another study [3] addresses the sampling of continuous-time periodic band-limited signals corrupted by additive noise. By modeling the noise as a stream of Dirac pulses, the study shows that the sum of a band-limited signal with a stream of Dirac pulses falls into the category of signals with a finite rate of innovation, implying a finite number of degrees of freedom. Additional works [4, 5] propose algorithms for perfect reconstruction of periodic band-limited signals from nonuniform samples, utilizing techniques such as Lagrange interpolation for trigonometric polynomials and iterative frame algorithms.

This monography focuses on the sampling of an analog multi-harmonic input signal, contrasting it with methods based on signal integration [6], along with the subsequent reconstruction of the processed signals. The approach in [6] employs standard matrix inversion for reconstruction, which requires computationally intensive numerical calculations. However, it has been observed in [7] that the ac signal integration method leads to a regular system matrix in the resulting system of equations,

P. Petrović, *Processing, Estimation and Measurement of Signals Parameters in Public Distribution Networks*, Studies in Systems, Decision and Control 502,
https://doi.org/10.1007/978-3-031-43107-4_3

provided appropriate choices are made regarding the integration time parameters. This observation enables more efficient reconstruction procedures.

While the presented algorithm is suitable for reconstructing periodic band-limited signals in simulations assuming error-free measurements, in practical environments, measurement errors occur. These errors arise due to imprecisions introduced by voltage and current transducers used for signal adaptation, conversion errors during analog-to-digital signal processing, and determination of the digital equivalent. In practice, the obtained samples may not represent the exact signal values at the sampling points but rather average values in their vicinity [5]. Consequently, modifications must be made to the suggested algorithm to obtain the best signal estimate based on the assumed criterion, similar to the approaches in [5, 6, 8–11].

The proposed algorithm [12] for estimation utilizes values obtained from continuous input signal sampling. This processing is performed iteratively until the unknown signal parameters can be determined. The resulting system of linear equations can be easily solved using derived analytical expressions, making the method computationally efficient compared to standard reconstruction algorithms with lower numerical error. The algorithm is specifically designed for highly accurate RMS measurements of periodic signals and can be extended to precise measurements of other important quantities such as power and energy.

3.1 Problem Formulation

Consider an input signal with a fundamental frequency f that is band-limited to the first M harmonic components. This continuous signal, characterized by its complex harmonic content, can be expressed as a summation of Fourier components, as presented in Table 3.1:

$$s(t) = a_0 + \sum_{k=1}^{M} a_k \sin(k2\pi f t + \psi_k) \tag{3.1}$$

By discretely sampling the signal (3.1) and constructing a system of equations with a similar structure, we can solve for the $2M + 1$ unknowns, which include the amplitudes and phases of the M harmonics along with the average value of the signal. The resulting system can be expressed as:

$$s(t_l) = a_0 + \sum_{k=1}^{M} a_k \sin(k2\pi f t_l + \psi_k) \tag{3.2}$$

Table 3.1 The nomenclature used in the following parts of the chapter

Symbol	Meaning
f	Fundamental frequency of processing signal
$s(t)$	Band limited input analogue signal
M	The number of signal spectrum components
ADC	Analogue to digital converter
a_0	The average value of the input signal
a_k	The amplitude of the *kth* harmonic
k	The number of the harmonic
ψ_k	The phase angle of the *kth* harmonic

In the above equation, where l ranges from 1 to $2M+1$, t_l denotes the time moment at which the input analog signal is sampled, and $s(t_l)$ represents the corresponding value of the processed signal at the sampling instant. This relation can be succinctly expressed as:

$$a_0 + \sum_{k=1}^{M} a_k\left(\sin\alpha_{k,l}\cos\psi_k + \cos\alpha_{k,l}\sin\psi_k\right) = s(t_l) \tag{3.3}$$

where:

$$\alpha_{k,l} = 2k\pi f t_l = k\alpha_l; \quad (k = 1, 2, \ldots, M); (l = 1, 2, \ldots, 2M+1) \tag{3.4}$$

The variables $\alpha_{k,l}$ in Eq. (3.3) are determined by the sampling moment and the frequency of the corresponding harmonic in the input periodic signal. The determinant of the system for the $2M+1$ unknown parameters in Eq. (3.3) can be denoted as:

$$\mathbf{X}_{system} = \begin{vmatrix} a_0 & a_1\sin\alpha_{1,1} & \ldots & a_M\sin\alpha_{M,1} & a_1\cos\alpha_{1,1} & \ldots & a_M\cos\alpha_{M,1} \\ a_0 & a_1\sin\alpha_{1,2} & \ldots & a_M\sin\alpha_{M,2} & a_1\cos\alpha_{1,2} & \ldots & a_M\cos\alpha_{M,2} \\ . & & \ldots. & . & . & \ldots. & . \\ a_0 & a_1\sin\alpha_{1,2M+1} & \ldots & a_M\sin\alpha_{M,2M+1} & a_1\cos\alpha_{1,2M+1} & \ldots & a_M\cos\alpha_{M,2M+1} \end{vmatrix} =$$
$$= a_0\left(a_1a_2\cdot\ldots\cdot a_{M-1}a_M\right)^2\mathbf{X}_{2M=1} \tag{3.5}$$

where:

$$\mathbf{X}_{2M+1} = \begin{vmatrix} 1 & \sin\alpha_{1,1} & \ldots & \sin\alpha_{M,1} & \cos\alpha_{1,1} & \ldots & \cos\alpha_{M,1} \\ 1 & \sin\alpha_{1,2} & \ldots & \sin\alpha_{M,2} & \cos\alpha_{1,2} & \ldots & \cos\alpha_{M,2} \\ . & . & \ldots. & & & \ldots & \\ 1 & \sin\alpha_{1,2M+1} & \ldots & \sin\alpha_{M,2M+1} & \cos\alpha_{1,2M+1} & \ldots & \cos\alpha_{M,2M+1} \end{vmatrix} =$$

$$= \begin{vmatrix} 1 & \sin\alpha_1 & \ldots & \sin M\alpha_1 & \cos\alpha_1 & \ldots & \cos M\alpha_1 \\ 1 & \sin\alpha_2 & \ldots & \sin M\alpha_2 & \cos\alpha_2 & \ldots & \cos M\alpha_2 \\ . & . & \ldots. & & & \ldots & \\ 1 & \sin\alpha_{2M+1} & \ldots & \sin M\alpha_{2M+1} & \cos\alpha_{2M+1} & \ldots & \cos M\alpha_{2M+1} \end{vmatrix} \quad (3.6)$$

A similar determinant to the one obtained in Eq. (3.6) has been studied in [7]. In its form, this determinant bears resemblance to the well-known Van der Monde determinant, which has been extensively analyzed in previous works [13–16]. However, the existing papers mainly proposed program-based procedures, often iterative in nature, for efficiently determining the value of the original and inverted Van der Monde matrix. In contrast, this paper presents novel analytical and concise expressions for solving the system of equations (3.3) by leveraging the aforementioned determinant as the starting point of the analysis. With the derived relationships, the standard procedure for solving the system of equations, as suggested in [6], becomes unnecessary. This is particularly advantageous when dealing with signals of extremely complex spectral content, where the standard procedure would demand substantial computational resources and time for processing.

3.1.1 The Determinants of the Van der Monde Matrix

Let us now examine the Mth order Van der Monde determinant, as described in previous works [15, 16]:

$$D_M = \Delta_M = \begin{vmatrix} 1 & x_1 & x_1^2 & \ldots & x_1^{M-1} \\ 1 & x_2 & x_2^2 & \ldots & x_2^{M-1} \\ \ldots & \ldots & \ldots & \ldots & \ldots \\ 1 & x_M & x_M^2 & \ldots & x_M^{M-1} \end{vmatrix} = \prod_{k=j+1}^{M} \prod_{j=1}^{M-1} (x_k - x_j) \quad (3.7)$$

If we introduce the following notation:

$$D_{i,j} = cof(D_M);\ D_{i,j} = (-1)^{i+j}\Delta_{i,j} \quad (3.8)$$

where $\Delta_{i,j}$ is the cofactor (minor) which is obtained from determinant Δ_M after the i row and the j column have been eliminated. The expansion of Δ_M in the last row yields:

$$\Delta_M = 1 \cdot D_{M,1} + x_M D_{M,2} + \ldots + x_M^{r-1} D_{M,r} + \ldots + x_M^{M-1} D_{M,M} \tag{3.9}$$

On the contrary, we have the following expression:

$$\Delta_M = (x_M - x_1)(x_M - x_2)\ldots(x_M - x_{M-1}) \prod_{k=j+1}^{M-1} \prod_{j=1}^{M-2} (x_k - x_j) \tag{3.10}$$

$$D_{M,M} = \prod_{k=j+1}^{M-1} \prod_{j=1}^{M-2} (x_k - x_j) \tag{3.11}$$

$$\Delta_M = (x_M - x_1)(x_M - x_2)\ldots(x_M - x_{M-1}) D_{M,M} \tag{3.12}$$

$D_{M,M}$ is the coefficient along with x_M^{M-1} in developing Δ_M from the degrees of x_M. This yield:

$$D_{M,M-1} = -(x_1 + x_2 + \ldots + x_{M-1}) D_{M,M} \tag{3.13}$$

$$D_{M,M-2} = (-1)^2 \sum (x_1 + x_2 + \ldots + x_{M-1})_2 D_{M,M} \tag{3.14}$$

$$D_{M,r} = (-1)^{M-r} \sum (x_1 + x_2 + \ldots + x_{M-1})_{M-r} D_{M,M} \tag{3.15}$$

$$D_{M,1} = (-1)^{M-1} (x_1 x_2 \ldots x_{M-1}) D_{M,M} \tag{3.16}$$

This infers that:

$$D_{M,r} = (-1)^{M-r} (x_1 x_2 \ldots x_{M-1}) \sum \frac{1}{(x_1 x_2 \ldots x_{M-1})_{r-1}} D_{M,M} \tag{3.17}$$

$$\Delta_{M,r} = (x_1 x_2 \ldots x_{M-1}) \sum \frac{1}{(x_1 x_2 \ldots x_{M-1})_{r-1}} D_{M,r} \tag{3.18}$$

Now, we can determine $\Delta_{p,q}$ and $D_{p,q}$:

$$\Delta_M = \begin{vmatrix} 1 & x_1 & \dots & x_1^{q-1} & \dots & x_1^{M-1} \\ \dots & \dots & \dots & \dots & \dots & \dots \\ 1 & x_p & \dots & x_p^{q-1} & \dots & x_p^{M-1} \\ \dots & \dots & \dots & \dots & \dots & \dots \\ 1 & x_M & \dots & x_M^{q-1} & \dots & x_M^{M-1} \end{vmatrix} = (-1)^{n-p} \begin{vmatrix} 1 & x_1 & \dots & x_1^{q-1} & \dots & x_1^{M-1} \\ \dots & \dots & \dots & \dots & \dots & \dots \\ 1 & x_{p-1} & \dots & x_{p-1}^{q-1} & \dots & x_{p-1}^{M-1} \\ 1 & x_{p+1} & \dots & x_{p+1}^{q-1} & \dots & x_{p+1}^{M-1} \\ \dots & \dots & \dots & \dots & \dots & \dots \\ 1 & x_M & \dots & x_M^{q-1} & \dots & x_M^{M-1} \\ 1 & x_p & \dots & x_p^{q-1} & \dots & x_p^{M-1} \end{vmatrix} \Rightarrow$$

$$\begin{aligned} \Delta_{p,q} &= (-1)^{M-p}\left(x_1 \dots x_{p-1}x_{p+1} \dots x_M\right) \cdot \\ &\cdot \sum \frac{1}{\left(x_1 \dots x_{p-1}x_{p+1} \dots x_M\right)_{q-1}} \cdot \\ &\cdot \frac{\Delta_M}{\left(x_M - x_p\right)\left(x_{M-1} - x_p\right) \dots \left(x_{p+1} - x_p\right)\left(x_p - x_{p-1}\right) \dots \left(x_p - x_1\right)} \end{aligned} \tag{3.19}$$

It follows that:

$$\begin{aligned} \Delta_{p,q} &= \frac{\Delta_M}{\left(x_p - x_1\right)\left(x_p - x_2\right) \dots \left(x_p - x_{p-1}\right)\left(x_p - x_{p+1}\right) \dots \left(x_p - x_M\right)} \cdot \\ &\cdot \left(x_1 \dots x_{p-1}x_{p+1} \dots x_M\right) \sum \frac{1}{\left(x_1 \dots x_{p-1}x_{p+1} \dots x_M\right)_{q-1}} \\ D_{p,q} &= (-1)^{p+q} \Delta_{p,q} \end{aligned} \tag{3.20}$$

(the sum of all inverted $(q-1)$ products of different indices).

3.1.2 *Derivation of the New Relations for Solving of the Observed System of Equations*

In the case of the system of equations formed according to the proposed signal processing concept (Eq. (3.3)), we replace the x variables in the above expressions with the trigonometric values for $\alpha_1, \alpha_2, \alpha_3, \dots, \alpha_{2M+1}$ as defined in Eq. (3.4). By utilizing *Euler's* formulas, we can transform the given determinant (Eq. (3.6)) as follows:

$$
\begin{aligned}
\mathbf{X}_{2M+1} &= \frac{1}{2^{2M}} e^{-M\frac{\pi}{2}i} \begin{vmatrix} 1 & e^{\alpha_1 i} - e^{-\alpha_1 i} & \dots & e^{M\alpha_1 i} - e^{-M\alpha_1 i} & e^{\alpha_1 i} + e^{-\alpha_1 i} & \dots & e^{M\alpha_1 i} + e^{-M\alpha_1 i} \\ 1 & e^{\alpha_2 i} - e^{-\alpha_2 i} & \dots & e^{M\alpha_2 i} - e^{-M\alpha_2 i} & e^{\alpha_2 i} + e^{-\alpha_2 i} & \dots & e^{M\alpha_2 i} + e^{-M\alpha_2 i} \\ . & . & \dots & & & \dots & \\ 1 & e^{\alpha_{2M+1} i} - e^{-\alpha_{2M+1} i} & \dots & e^{M\alpha_{2M+1} i} - e^{-M\alpha_{2M+1} i} & e^{\alpha_{2M+1} i} + e^{-\alpha_{2M+1} i} & \dots & e^{M\alpha_{2M+1} i} + e^{-M\alpha_{2M+1} i} \end{vmatrix} \\
&= \frac{1}{2^{2M}} e^{-M\frac{\pi}{2}i} \left| 1 \; e^{\alpha_1 i} - e^{-\alpha_1 i} \; e^{2\alpha_1 i} - e^{-2\alpha_1 i} \; \dots \; e^{M\alpha_1 i} - e^{-M\alpha_1 i} \; e^{\alpha_1 i} + e^{-\alpha_1 i} \; e^{2\alpha_1 i} + e^{-2\alpha_1 i} \; \dots \; e^{M\alpha_1 i} + e^{-M\alpha_1 i} \right| = \\
&= \frac{1}{2^{2M}} e^{-M\frac{\pi}{2}i} \left| 1 \; e^{\alpha_1 i} - e^{-\alpha_1 i} \; e^{2\alpha_1 i} - e^{-2\alpha_1 i} \; \dots \; e^{M\alpha_1 i} - e^{-M\alpha_1 i} \; e^{\alpha_1 i} \; e^{2\alpha_1 i} \; \dots \; e^{M\alpha_1 i} \right| = \\
&= \frac{(-1)^M}{2^{2M}} e^{-M\frac{\pi}{2}i} \left| 1 \; e^{-\alpha_1 i} \; e^{-2\alpha_1 i} \; \dots \; e^{-M\alpha_1 i} \; e^{\alpha_1 i} \; e^{2\alpha_1 i} \; \dots \; e^{M\alpha_1 i} \right| = \\
&= \frac{(-1)^{\frac{M(M-1)}{2}}}{2^{2M}} e^{-M\frac{\pi}{2}i} \left| e^{-M\alpha_1 i} \; e^{-(M-1)\alpha_1 i} \; \dots \; e^{-\alpha_1 i} \; 1 \; e^{\alpha_1 i} \; e^{2\alpha_1 i} \; \dots \; e^{M\alpha_1 i} \right| = \\
&= \frac{(-1)^{\frac{M(M-1)}{2}}}{2^{2M}} e^{-M\frac{\pi}{2}i} e^{-M(\alpha_1 + \dots + \alpha_{2M+1})i} \left| 1 \; e^{\alpha_1 i} \; e^{2\alpha_1 i} \; \dots \; e^{(M-1)\alpha_1 i} \; e^{M\alpha_1 i} \; e^{(M+1)\alpha_1 i} \; \dots \; e^{2M\alpha_1 i} \right| = \\
&= \frac{(-1)^{\frac{M(M-1)}{2}}}{2^{2M}} e^{-M\frac{\pi}{2}i} e^{-M(\alpha_1 + \dots + \alpha_{2M+1})i} \prod_{j=k+1}^{2M+1} \prod_{k=1}^{2M} \left(e^{\alpha_j i} - e^{\alpha_k i} \right) = \\
&= \frac{(-1)^{\frac{M(M+1)}{2}} i^M}{2^M} e^{-M(\alpha_1 + \dots + \alpha_{2M+1})i} \prod_{j=k+1}^{2M+1} \prod_{k=1}^{2M} \left(e^{\frac{\alpha_j + \alpha_k}{2} i} 2i \sin \frac{\alpha_j - \alpha_k}{2} \right) = \\
&= \frac{(-1)^{\frac{M(M+1)}{2}} i^M}{2^M} e^{-M(\alpha_1 + \dots + \alpha_{2M+1})i} e^{M(\alpha_1 + \dots + \alpha_{2M+1})i} 2^{\frac{2M(2M+1)}{2}} i^{\frac{2M(2M+1)}{2}} \prod_{j=k+1}^{2M+1} \prod_{k=1}^{2M} \sin \frac{\alpha_j - \alpha_k}{2} \Rightarrow
\end{aligned}
$$

$$
\mathbf{X}_{2M+1} = (-1)^{\frac{M(M+1)}{2}} 2^{2M^2} \prod_{j=k+1}^{2M+1} \prod_{k=1}^{2M} \sin \frac{\alpha_j - \alpha_k}{2} \tag{3.21}
$$

The cofactors that correspond to the observed system of equations (3.3) are:

$$\mathbf{X}_{2M+1,1} = \begin{vmatrix} s(t_1) & \sin\alpha_1 & \dots & \sin M\alpha_1 & \cos\alpha_1 & \dots & \cos M\alpha_1 \\ s(t_2) & \sin\alpha_2 & \dots & \sin M\alpha_2 & \cos\alpha_2 & \dots & \cos M\alpha_2 \\ . & . & \dots & & & \dots & \\ s(t_{2M+1}) & \sin\alpha_{2M+1} & \dots & \sin M\alpha_{2M+1} & \cos\alpha_{2M+1} & \dots & \cos M\alpha_{2M+1} \end{vmatrix} \tag{3.22}$$

$$\mathbf{X}_{2M+1,2} = \begin{vmatrix} 1 & s(t_1) & \sin 2\alpha_1 & \dots & \sin M\alpha_1 & \cos\alpha_1 & \cos 2\alpha_1 & \dots & \cos M\alpha_1 \\ 1 & s(t_2) & \sin 2\alpha_2 & \dots & \sin M\alpha_2 & \cos\alpha_2 & \cos 2\alpha_2 & \dots & \cos M\alpha_2 \\ . & . & . & & & & & & \\ 1 & s(t_{2M+1}) & \sin 2\alpha_{2M+1} & \dots & \sin M\alpha_{2M+1} & \cos\alpha_{2M+1} & \cos 2\alpha_{2M+1} & \dots & \cos M\alpha_{2M+1} \end{vmatrix} \tag{3.23}$$

and so on. The cofactors given above based on the following development can be written as:

$$\mathbf{X}_{2M+1,1} = s(t_1)\mathbf{X}_1^1 + s(t_2)\mathbf{X}_2^1 + \dots + s(t_{2M+1})\mathbf{X}_{2M+1}^1 \tag{3.24}$$

The variables $\mathbf{X}_1^1, \mathbf{X}_2^1, \dots, \mathbf{X}_{2M+1}^1$ represent the cofactors obtained from the cofactor $\mathbf{X}_{2M+1,1}$ after eliminating the corresponding row and the first column. The second cofactor, denoted as $\mathbf{X}_{2M+1,2}$ is derived from expanding $\mathbf{X}_{2M+1}$ along a specific column. To determine these cofactors, we need to calculate $\mathbf{X}_p^q$ 1 for $1 \leq q \leq M+1$. Therefore, we have the following expression:

$$\mathbf{X}_p^q = (-1)^{p+q}\mathbf{E}_p^q \tag{3.25}$$

where $\mathbf{E}_p^q$ is obtained from $\mathbf{X}_{2M+I}$ overturning p row and q column. We know that: $\mathbf{X}_{2M+1} = \mathbf{E}_{2M+1}$.

$$\mathbf{E}_p^1 = \frac{e^{-M\frac{\pi}{2}i}}{2^{2M}} \left| e^{\alpha_1 i} - e^{-\alpha_1 i} \ \dots \ e^{M\alpha_1 i} - e^{-M\alpha_1 i} \ \ e^{\alpha_1 i} + e^{-\alpha_1 i} \ \ e^{2\alpha_1 i} + e^{-2\alpha_1 i} \ \dots \ e^{M\alpha_1 i} + e^{-M\alpha_1 i} \right|_p \tag{3.26}$$

The index p shows that p row is eliminated from this determinant.

$$\begin{aligned} \mathbf{E}_p^1 &= \frac{e^{-M\frac{\pi}{2}i}}{2^{2M}} \left| e^{\alpha_1 i} - e^{-\alpha_1 i} \ \dots \ e^{M\alpha_1 i} - e^{-M\alpha_1 i} \ \ e^{\alpha_1 i} + e^{-\alpha_1 i} \ \ e^{2\alpha_1 i} + e^{-2\alpha_1 i} \ \dots \ e^{M\alpha_1 i} + e^{-M\alpha_1 i} \right|_p = \\ &= \frac{(-1)^M}{2^M} e^{-M\frac{\pi}{2}i} \left| e^{-\alpha_1 i} \ \dots \ e^{-M\alpha_1 i} \ \ e^{\alpha_1 i} \ \ e^{2\alpha_1 i} \ \dots \ e^{M\alpha_1 i} \right|_p = \\ &= \frac{(-1)^{\frac{M(M+1)}{2}}}{2^M} e^{-M\frac{\pi}{2}i} \left| e^{-M\alpha_1 i} \ \dots \ e^{-\alpha_1 i} \ \ e^{\alpha_1 i} \ \dots \ e^{M\alpha_1 i} \right|_p = \\ &= \frac{(-1)^{\frac{M(M+1)}{2}}}{2^M} e^{-M\frac{\pi}{2}i} e^{-M(\alpha_1+\dots+\alpha_{p-1}+\alpha_{p+1}+\dots+\alpha_{2M+1})i} \left| 1 \ \ e^{\alpha_1 i} \ \dots \ e^{(M-1)\alpha_1 i} \ \ e^{(M+1)\alpha_1 i} \ \dots \ e^{2M\alpha_1 i} \right|_p = \\ &= \frac{(-1)^{\frac{M(M+1)}{2}}}{2^M} e^{-M\frac{\pi}{2}i} e^{-M(\alpha_1+\dots+\alpha_{p-1}+\alpha_{p+1}+\dots+\alpha_{2M+1})i} \Delta_{p,M+1}\left(x_j = e^{\alpha_j i}\right) \end{aligned} \tag{3.27}$$

$$\Delta_{p,M+1}\left(x_j = e^{\alpha_j i}\right) =$$

$$= \frac{\Delta_{2M+1}}{\left(x_p - x_1\right)\left(x_p - x_2\right)\ldots\left(x_p - x_{p-1}\right)\left(x_p - x_{p+1}\right)\ldots\left(x_p - x_{2M+1}\right)}\cdot$$

$$\cdot\left(x_1 \ldots x_{p-1}x_{p+1} \ldots x_{2M+1}\right)\sum \frac{1}{\left(x_1 \ldots x_{p-1}x_{p+1} \ldots x_{2M+1}\right)_M}\Bigg|_{x_j = e^{\alpha_j i}} \tag{3.28}$$

$$\Delta_{2M+1}\left(x_j = e^{\alpha_j i}\right) =$$

$$= 2^{M(2M+1)}e^{M(2M+1)\frac{\pi}{2}i}e^{M(\alpha_1+\alpha_2+\ldots+\alpha_{2M+1})i}\prod_{j=k+1}^{2M+1}\prod_{k=1}^{2M}\sin\frac{\alpha_j - \alpha_k}{2} \tag{3.29}$$

$$\left(x_p - x_1\right)\left(x_p - x_2\right)\ldots\left(x_p - x_{p-1}\right)\left(x_p - x_{p+1}\right)\ldots\left(x_p - x_{2M+1}\right)\Big|_{x_j = e^{\alpha_j i}} =$$

$$= (-1)^M 2^{2M} e^{M\alpha_p i} e^{\frac{1}{2}(\alpha_1+\ldots+\alpha_{p-1}+\alpha_{p+1}+\ldots+\alpha_{2M+1})i}\prod_{\substack{k=1\\k\neq p}}^{2M+1}\sin\frac{\alpha_p - \alpha_k}{2} \tag{3.30}$$

$$\Delta_{2M+1}\left(x_j = e^{\alpha_j i}\right) = (-1)^M 2^{M(2M+1)}e^{M\frac{\pi}{2}i}e^{M(\alpha_1+\alpha_2+\ldots+\alpha_{2M+1})i}\cdot$$

$$\cdot\prod_{j=k+1}^{2M+1}\prod_{k=1}^{2M}\sin\frac{\alpha_j - \alpha_k}{2}(-1)^M 2^{-2M}e^{-M\alpha_p i}e^{-\frac{1}{2}(\alpha_1+\ldots+\alpha_{p-1}+\alpha_{p+1}+\ldots+\alpha_{2M+1})i}\cdot$$

$$\cdot\frac{1}{\prod\limits_{\substack{k=1\\k\neq p}}^{2M+1}\sin\frac{\alpha_p-\alpha_k}{2}}\cdot e^{(\alpha_1+\ldots+\alpha_{p-1}+\alpha_{p+1}+\ldots+\alpha_{2M+1})i}\cdot$$

$$\cdot\sum e^{-(\alpha_1+\ldots+\alpha_{p-1}+\alpha_{p+1}+\ldots+\alpha_{2M+1})i}|_M \Rightarrow$$

$$\mathbf{E}_p^1 = (-1)^{\frac{M(M+1)}{2}}2^{2M(M-1)}\frac{\prod\limits_{j=k+1}^{2M+1}\prod\limits_{k=1}^{2M}\sin\frac{\alpha_j-\alpha_k}{2}}{\prod\limits_{\substack{k=1\\k\neq p}}^{2M+1}\sin\frac{\alpha_p-\alpha_k}{2}}\cdot e^{\frac{1}{2}(\alpha_1+\ldots+\alpha_{p-1}+\alpha_{p+1}+\ldots+\alpha_{2M+1})i}\cdot$$

$$\cdot\sum e^{-(\alpha_1+\ldots+\alpha_{p-1}+\alpha_{p+1}+\ldots+\alpha_{2M+1})i}|_M \tag{3.31}$$

or:

$$\mathbf{E}_p^1 = (-1)^{\frac{M(M+1)}{2}} 2^{2M(M-1)} \frac{\prod\limits_{j=k+1}^{2M+1} \prod\limits_{k=1}^{2M} \sin \frac{\alpha_j - \alpha_k}{2}}{\prod\limits_{\substack{k=1 \\ k \neq p}}^{2M+1} \sin \frac{\alpha_p - \alpha_k}{2}} \cdot \\ \cdot \sum \cos \left\{ \begin{array}{l} \frac{\alpha_1 + \ldots + \alpha_{p-1} + \alpha_{p+1} + \ldots + \alpha_{2M+1}}{2} - \\ -\left(\alpha_1 + \ldots + \alpha_{p-1} + \alpha_{p+1} + \ldots + \alpha_{2M+1}\right)_M \end{array} \right\} \tag{3.32}$$

It follows that:

$$\frac{\mathbf{X}_p^1}{\mathbf{X}_{2M+1}} = \frac{(-1)^{p+1}}{2^{2M}} \frac{1}{\prod\limits_{1 \leq k \neq p}^{2M+1} \sin \frac{\alpha_p - \alpha_k}{2}} \cdot \\ \cdot \sum \cos \left\{ \begin{array}{l} \frac{\alpha_1 + \ldots + \alpha_{p-1} + \alpha_{p+1} + \ldots + \alpha_{2M+1}}{2} - \\ -\left(\alpha_1 + \ldots + \alpha_{p-1} + \alpha_{p+1} + \ldots + \alpha_{2M+1}\right)_M \end{array} \right\} \tag{3.33}$$

Now, we can determine $\mathbf{E}_p^q$ for $2 \leq q \leq M+1$:

$$
\begin{aligned}
\mathbf{E}_p^q &= \frac{e^{-(M-1)\frac{\pi}{2}i}}{2^{2M-1}} \left| 1 \; e^{\alpha_1 i} - e^{-\alpha_1 i} \; \dots \; e^{(q-2)\alpha_1 i} - e^{-(q-2)\alpha_1 i} \; e^{q\alpha_1 i} - e^{-q\alpha_1 i} \; \dots \; e^{M\alpha_1 i} - e^{-M\alpha_1 i} \; e^{\alpha_1 i} + e^{-\alpha_1 i} \; \dots \; e^{M\alpha_1 i} + e^{-M\alpha_1 i} \right|_p = \\
&= \frac{(-1)^{M-1}}{2^M} e^{-(M-1)\frac{\pi}{2}i} \left\{ \begin{array}{l} \left| 1 \; e^{-\alpha_1 i} \; \dots \; e^{-(q-2)\alpha_1 i} \; e^{-q\alpha_1 i} \; \dots \; e^{-M\alpha_1 i} \; e^{\alpha_1 i} \; \dots \; e^{M\alpha_1 i} \right|_p + \\ +(-1)^{M-1} \left| 1 \; e^{-\alpha_1 i} \; \dots \; e^{-M\alpha_1 i} \; e^{\alpha_1 i} \; \dots \; e^{(q-2)\alpha_1 i} \; e^{q\alpha_1 i} \; \dots \; e^{M\alpha_1 i} \right|_p \end{array} \right\} = \\
&= \frac{(-1)^{M-1}}{2^M} e^{-(M-1)\frac{\pi}{2}i} \left\{ \begin{array}{l} (-1)^{\frac{M(M-1)}{2}} \left| e^{-M\alpha_1 i} \; \dots \; e^{-q\alpha_1 i} \; e^{-(q-2)\alpha_1 i} \; \dots \; e^{-\alpha_1 i} \; 1 \; e^{\alpha_1 i} \; \dots \; e^{M\alpha_1 i} \right|_p + \\ +(-1)^{M-1}(-1)^{\frac{M(M-1)}{2}} \left| e^{-M\alpha_1 i} \; \dots \; e^{-\alpha_1 i} \; 1 \; e^{\alpha_1 i} \; \dots \; e^{(q-2)\alpha_1 i} \; e^{q\alpha_1 i} \; \dots \; e^{M\alpha_1 i} \right|_p \end{array} \right\} = \\
&= \frac{(-1)^{\frac{M(M+1)}{2}}}{2^M} e^{-(M-1)\frac{\pi}{2}i} e^{-M\left(\alpha_1 + \dots + \alpha_{p-1} + \alpha_{p+1} + \dots + \alpha_{2M+1}\right)i} \left\{ \begin{array}{l} \left| 1 \; \dots \; e^{(M-q)\alpha_1 i} \; e^{(M-q+2)\alpha_1 i} \; \dots \; e^{(M-1)\alpha_1 i} \; \dots \; e^{2M\alpha_1 i} \right|_p - \\ - \left| 1 \; \dots \; e^{(M-1)\alpha_1 i} \; e^{M\alpha_1 i} \; \dots \; e^{(M+q-2)\alpha_1 i} \; e^{(M+q)\alpha_1 i} \; \dots \; e^{2M\alpha_1 i} \right|_p \end{array} \right\} = \\
&= \frac{(-1)^{\frac{M(M+1)}{2}}}{2^M} e^{-(M-1)\frac{\pi}{2}i} e^{-M\left(\alpha_1 + \dots + \alpha_{p-1} + \alpha_{p+1} + \dots + \alpha_{2M+1}\right)i} \left\{ \left(\Delta_{p,M-q+2} - \Delta_{p,M+q} \right) \left(x_j = e^{\alpha_j i} \right) \right\}
\end{aligned}
\tag{3.34}
$$

Now is:

$$\left(\Delta_{p,M-q+2}-\Delta_{p,M+q}\right)\left(x_j=e^{\alpha_j i}\right)=\frac{\Delta_{2M+1}}{\prod\limits_{1\le k\ne p}^{2M+1}\left(x_p-x_k\right)}\cdot$$

$$\cdot\left(x_1x_2\ldots x_{p-1}x_{p+1}\ldots x_{2M+1}\right)$$

$$\cdot\left\{\begin{array}{l}\sum\frac{1}{\left(x_1x_2\ldots x_{p-1}x_{p+1}\ldots x_{2M+1}\right)_{M-q+1}}-\\-\sum\frac{1}{\left(x_1x_2\ldots x_{p-1}x_{p+1}\ldots x_{2M+1}\right)_{M+q-1}}\end{array}\right\}\Bigg|_{x_j=e^{\alpha_j i}}\Rightarrow$$

$$\mathbf{E}_p^q=(-1)^{\frac{M(M+1)}{2}}2^{2M(M-1)+1}\frac{\prod\limits_{j=k+1}^{2M+1}\prod\limits_{k=1}^{2M}\sin\frac{\alpha_j-\alpha_k}{2}}{\prod\limits_{1\le k\ne p}^{2M+1}\sin\frac{\alpha_p-\alpha_k}{2}}$$

$$\cdot\sum\sin\left\{\begin{array}{l}\frac{\alpha_1+\ldots+\alpha_{p-1}+\alpha_{p+1}+\ldots+\alpha_{2M+1}}{2}-\\-\left(\alpha_1+\ldots+\alpha_{p-1}+\alpha_{p+1}+\ldots+\alpha_{2M+1}\right)_{M+q-1}\end{array}\right\}\tag{3.35}$$

It follows that:

$$\frac{\mathbf{X}_p^q}{\mathbf{X}_{2M+1}}=\frac{(-1)^{p+q}}{2^{2M-1}}\frac{1}{\prod\limits_{1\le k\ne p}^{2M+1}\sin\frac{\alpha_p-\alpha_k}{2}}\cdot$$

$$\cdot\sum\sin\left\{\begin{array}{l}\frac{\alpha_1+\ldots+\alpha_{p-1}+\alpha_{p+1}+\ldots+\alpha_{2M+1}}{2}-\\-\left(\alpha_1+\ldots+\alpha_{p-1}+\alpha_{p+1}+\ldots+\alpha_{2M+1}\right)_{M+q-1}\end{array}\right\};2\le q\le M\tag{3.36}$$

while:

$$\frac{\mathbf{X}_p^q}{\mathbf{X}_{2M+1}}=\frac{(-1)^{p+q}}{2^{2M-1}}\frac{1}{\prod\limits_{1\le k\ne p}^{2M+1}\sin\frac{\alpha_p-\alpha_k}{2}}\cdot$$

$$\cdot\sin\frac{\alpha_1+\ldots+\alpha_{p-1}+\alpha_{p+1}+\ldots+\alpha_{2M+1}}{2};\ for\ q=M+1\tag{3.37}$$

For $q=M+r+1$:

$$
\begin{aligned}
\mathbf{E}_p^q &= \left|1 \;\; \sin\alpha_1 \;\; \ldots \;\; \sin M\alpha_1 \;\; \cos\alpha_1 \;\; \ldots \;\; \cos(r-1)\alpha_1 \;\; \cos(r+1)\alpha_1 \;\; \ldots \;\; \cos M\alpha_1\right|_p \\
&= \frac{e^{-M\frac{\pi}{2}i}}{2^{2M-1}}\left|1 \;\; e^{\alpha_1 i}-e^{-\alpha_1 i} \;\; \ldots \;\; e^{M\alpha_1 i}-e^{-M\alpha_1 i} \;\; e^{\alpha_1 i}+e^{-\alpha_1 i} \;\; \ldots \;\; e^{(r-1)\alpha_1 i}+e^{-(r-1)\alpha_1 i} \;\; e^{(r+1)\alpha_1 i}+e^{-(r+1)\alpha_1 i} \;\; \ldots \;\; e^{M\alpha_1 i}+e^{-M\alpha_1 i}\right|_p \\
&= \frac{e^{-M\frac{\pi}{2}i}}{2^{M}}\left|1 \;\; e^{\alpha_1 i}-e^{-\alpha_1 i} \;\; \ldots \;\; e^{r\alpha_1 i}-e^{-r\alpha_1 i} \;\; \ldots \;\; e^{M\alpha_1 i}-e^{-M\alpha_1 i} \;\; e^{\alpha_1 i} \;\; \ldots \;\; e^{(r-1)\alpha_1 i} \;\; e^{(r+1)\alpha_1 i} \;\; \ldots \;\; e^{M\alpha_1 i}\right|_p \\
&= \frac{(-1)^{M-1}}{2^{M}}e^{-M\frac{\pi}{2}i}\left\|1 \;\; e^{-\alpha_1 i} \;\; \ldots \;\; e^{-(r-1)\alpha_1 i} \;\; e^{r\alpha_1 i}-e^{-r\alpha_1 i} \;\; e^{-(r+1)\alpha_1 i} \;\; \ldots \;\; e^{-M\alpha_1 i} \;\; e^{\alpha_1 i} \;\; \ldots \;\; e^{(r-1)\alpha_1 i} \;\; e^{(r+1)\alpha_1 i} \;\; \ldots \;\; e^{M\alpha_1 i}\right|_p \\
&= \frac{(-1)^{M-1}}{2^{M}}e^{-M\frac{\pi}{2}i}\left\{\begin{array}{l}(-1)^{M-1}\left|1 \;\; e^{-\alpha_1 i} \;\; \ldots \;\; e^{-(r-1)\alpha_1 i} \;\; e^{-(r+1)\alpha_1 i} \;\; \ldots \;\; e^{-M\alpha_1 i} \;\; e^{\alpha_1 i} \;\; \ldots \;\; e^{M\alpha_1 i}\right|_p \\ -\left|1 \;\; e^{-\alpha_1 i} \;\; \ldots \;\; e^{-M\alpha_1 i} \;\; e^{\alpha_1 i} \;\; \ldots \;\; e^{-(r-1)\alpha_1 i} \;\; e^{(r+1)\alpha_1 i} \;\; \ldots \;\; e^{M\alpha_1 i}\right|_p\end{array}\right\} \\
&= \frac{(-1)^{M-1}}{2^{M}}e^{-M\frac{\pi}{2}i}\left\{\begin{array}{l}(-1)^{M-1}(-1)^{\frac{M(M-1)}{2}}\left|e^{-M\alpha_1 i} \;\; \ldots \;\; e^{-(r+1)\alpha_1 i} \;\; e^{-(r-1)\alpha_1 i} \;\; \ldots \;\; e^{-\alpha_1 i} \;\; 1 \;\; e^{\alpha_1 i} \;\; \ldots \;\; e^{M\alpha_1 i}\right|_p - \\ -(-1)^{\frac{M(M-1)}{2}}\left|e^{-M\alpha_1 i} \;\; \ldots \;\; e^{-\alpha_1 i} \;\; 1 \;\; e^{\alpha_1 i} \;\; \ldots \;\; e^{-(r-1)\alpha_1 i} \;\; e^{(r+1)\alpha_1 i} \;\; \ldots \;\; e^{M\alpha_1 i}\right|_p\end{array}\right\} \\
&= -\frac{(-1)^{M-1}}{2^{M}}e^{-M\frac{\pi}{2}i}(-1)^{\frac{M(M-1)}{2}}e^{M\left(\alpha_1+\ldots+\alpha_{p-1}+\alpha_{p+1}+\ldots+\alpha_{2M+1}\right)i}\left(\Delta_{p,M-r+1}+\Delta_{p,M+r+1}\right) \Rightarrow
\end{aligned}
$$

$q = M + r + 1 \wedge 1 \le r \le M - 1$

$$
\mathbf{E}_p^q = (-1)^{\frac{M(M-1)}{2}}2^{2M(M-1)+1}\frac{\prod\limits_{j=k+1}^{2M+1}\prod\limits_{k=1}^{2M}\sin\frac{\alpha_j-\alpha_k}{2}}{\prod\limits_{1\le k\ne p}^{2M+1}\sin\frac{\alpha_p-\alpha_k}{2}}\sum\cos\left\{\begin{array}{l}\frac{\alpha_1+\ldots+\alpha_{p-1}+\alpha_{p+1}+\ldots+\alpha_{2M+1}}{2}- \\ -\left(\alpha_1+\ldots+\alpha_{p-1}+\alpha_{p+1}+\ldots+\alpha_{2M+1}\right)_{M-r}\end{array}\right\}
$$

$q = 2M + 1$

$$
\mathbf{E}_p^q = (-1)^{\frac{M(M-1)}{2}}2^{2M(M-1)+1}\frac{\prod\limits_{j=k+1}^{2M+1}\prod\limits_{k=1}^{2M}\sin\frac{\alpha_j-\alpha_k}{2}}{\prod\limits_{1\le k\ne p}^{2M+1}\sin\frac{\alpha_p-\alpha_k}{2}}\cos\frac{\alpha_1+\ldots+\alpha_{p-1}+\alpha_{p+1}+\ldots+\alpha_{2M+1}}{2} \tag{3.38}
$$

It follows that:

$$\begin{gathered} for\ q = M + r + 1 \wedge 1 \leq r \leq M - 1 \\ \frac{\mathbf{X}_p^q}{\mathbf{X}_{2M+1}} = \frac{(-1)^{p+r+1}}{2^{2M-1}} \frac{1}{\prod\limits_{1 \leq k \neq p}^{2M+1} \sin \frac{\alpha_p - \alpha_k}{2}} \\ \sum \cos \left\{ \begin{array}{l} \frac{\alpha_1 + \ldots + \alpha_{p-1} + \alpha_{p+1} + \ldots + \alpha_{2M+1}}{2} - \\ -\left(\alpha_1 + \ldots + \alpha_{p-1} + \alpha_{p+1} + \ldots + \alpha_{2M+1}\right)_{M-r} \end{array} \right\} \end{gathered} \tag{3.39}$$

$$\begin{gathered} for\ q = 2M + 1 \\ \frac{\mathbf{X}_p^q}{\mathbf{X}_{2M+1}} = \frac{(-1)^{p+M+1}}{2^{2M-1}} \frac{1}{\prod\limits_{1 \leq k \neq p}^{2M+1} \sin \frac{\alpha_p - \alpha_k}{2}} \\ \cos \frac{\alpha_1 + \ldots + \alpha_{p-1} + \alpha_{p+1} + \ldots + \alpha_{2M+1}}{2} \end{gathered} \tag{3.40}$$

We obtain:

$$\begin{gathered} \mathbf{X}_p^1 = (-1)^{p+q} (-1)^{\frac{M(M+1)}{2}} 2^{2M(M-1)} \frac{\prod\limits_{j=k+1}^{2M+1} \prod\limits_{k=1}^{2M} \sin \frac{\alpha_j - \alpha_k}{2}}{\prod\limits_{\substack{k=1 \\ k \neq p}}^{2M+1} \sin \frac{\alpha_p - \alpha_k}{2}} \cdot \\ \cdot \sum \cos \left\{ \begin{array}{l} \frac{\alpha_1 + \ldots + \alpha_{p-1} + \alpha_{p+1} + \ldots + \alpha_{2M+1}}{2} - \\ -\left(\alpha_1 + \ldots + \alpha_{p-1} + \alpha_{p+1} + \ldots + \alpha_{2M+1}\right)_M \end{array} \right\} \end{gathered} \tag{3.41}$$

(the summing is done by all of the M*'s* of the set $\{\alpha_1, \alpha_2, \ldots, \alpha_{2M+1}\}$).

For $2 \leq q \leq M + 1$:

$$\begin{gathered} \mathbf{X}_p^q = (-1)^{p+q} (-1)^{\frac{M(M+1)}{2}} 2^{2M(M-1)+1} \frac{\prod\limits_{j=k+1}^{2M+1} \prod\limits_{k=1}^{2M} \sin \frac{\alpha_j - \alpha_k}{2}}{\prod\limits_{1 \leq k \neq p}^{2M+1} \sin \frac{\alpha_p - \alpha_k}{2}} \cdot \\ \cdot \sum \sin \left\{ \begin{array}{l} \frac{\alpha_1 + \ldots + \alpha_{p-1} + \alpha_{p+1} + \ldots + \alpha_{2M+1}}{2} - \\ -\left(\alpha_1 + \ldots + \alpha_{p-1} + \alpha_{p+1} + \ldots + \alpha_{2M+1}\right)_{M+q-1} \end{array} \right\} \end{gathered} \tag{3.42}$$

$$q = M + r + 1 \wedge 1 \leq r \leq M - 1$$

$$\mathbf{X}_p^q = (-1)^{\frac{M(M-1)}{2}} 2^{2M(M-1)+1} \frac{\prod_{j=k+1}^{2M+1} \prod_{k=1}^{2M} \sin \frac{\alpha_j - \alpha_k}{2}}{\prod_{1 \leq k \neq p}^{2M+1} \sin \frac{\alpha_p - \alpha_k}{2}}$$

$$\sum \cos \left\{ \begin{array}{l} \frac{\alpha_1 + \ldots + \alpha_{p-1} + \alpha_{p+1} + \ldots + \alpha_{2M+1}}{2} - \\ -\left(\alpha_1 + \ldots + \alpha_{p-1} + \alpha_{p+1} + \ldots + \alpha_{2M+1}\right)_{M-r} \end{array} \right\}$$

$$q = 2M + 1$$

$$\mathbf{X}_p^q = (-1)^{\frac{M(M-1)}{2}} 2^{2M(M-1)+1} \frac{\prod_{j=k+1}^{2M+1} \prod_{k=1}^{2M} \sin \frac{\alpha_j - \alpha_k}{2}}{\prod_{1 \leq k \neq p}^{2M+1} \sin \frac{\alpha_p - \alpha_k}{2}}$$

$$\cos \frac{\alpha_1 + \ldots + \alpha_{p-1} + \alpha_{p+1} + \ldots + \alpha_{2M+1}}{2} \tag{3.43}$$

Using the derived analytical relation, the unknown parameters of the signal, such as amplitude and phase, can be determined by simply dividing the expression representing the solution of the corresponding co-determinants by the expression representing the analytical solution to the system determinant. The determinants and co-determinants of the system (3.3) were initially analyzed to obtain explicit and concise analytical expressions that can be used for calculating the unknown signal parameters, particularly in real-time scenarios where the frequency spectrum of the input signals is known. It is worth noting that the system of equations (3.3) can be represented, after processing, by a specific form of determinant, known as the Van der Monde's determinant. This property allows for factoring and application of transformations that are applicable only to determinants.

3.1.3 *Verification of Derived Relations*

Table 3.2 provides a comparison between the results obtained using the derived relations for solving the observed system of equations (3.3) and the GEPP algorithm (Gaussian elimination with partial pivoting) implemented in the Matlab program package. The calculations are performed using IEEE standard double floating point arithmetic with unit round off $u \approx 1.1 \times 10^{-16}$. This comparison serves as a practical validation of the proposed algorithm in the scenario of ideal sampling, where no errors are present in the sampled values and the determination of the processed signal's frequency.

Table 3.2 Verification of the derived expression for solving a system of equations with which the reconstruction of the observed signal is done

	$\mathbf{X}_{2M+1}$	$\mathbf{X}_l^1$; $l = 1, \ldots, 15$
Proposed algorithm	96.2746562	77.3751539; 240.7369977; 585.7387744; 1048.6023899; 1601.6571515; 2122.6636989; 2515.8984526; 2655.5007221; 557.8121810; 2190.4815359; 1686.1064501; 1129.5098525; 644.9058063; 279.7307319; 94.0066154
GEPP algorithm	96.2746562	77.3751539; 240.7369977; 585.7387744; 1048.6023899; 1601.6571515; 2122.6636989; 2515.8984526; 2655.5007221; 557.8121810; 2190.4815359; 1686.1064501; 1129.5098525; 644.9058063; 279.7307319; 94.0066154

The values in columns separated by commas correspond to the solution for derived relations with different orders (taking that $M = 7$, $f = 50$ Hz, $t_l = 0.001$ s). This means that in the column for $\mathbf{X}_l^1$ values 77.3751539; 65.8192372; −96.0730198; 82.9370773; etc. correspond to $\mathbf{X}_1^1 = 77.3751539$; $\mathbf{X}_2^1 = 65.8192372$; $\mathbf{X}_3^1 = -96.0730198$; $\mathbf{X}_4^1 = 82.9370773$, respectively.

The results presented in Table 3.2 demonstrate that the solutions obtained using the derived relations are virtually indistinguishable from the commonly used procedure for solving linear equation systems. The difference between the obtained values was found to be negligible, with a maximum discrepancy of 1×10^{-14}.

3.2 Proposed Reconstruction Algorithm and Uncertainty Analysis

The proposed algorithm for signal reconstruction is visually represented in the form of a flowchart, as shown in Fig. 3.1.

To apply the proposed algorithm, prior knowledge or an assumed value of the highest harmonic component M in the spectrum of the processed signal is required. Various methods can be employed to estimate the frequency spectrum. For instance, in [17], two accurate frequency estimation algorithms based on the linear prediction approach were developed. The first algorithm minimizes the weighted least squares (WLS) cost function with a generalized unit-norm constraint, while the second method is a WLS estimator with a monic constraint. Both algorithms provide close frequency estimates with accuracy approaching the Cramér–Rao lower bound for white Gaussian noise. Another approach, presented in [18], proposes a modified parameter estimator based on a magnitude phase-locked loop principle, which improves tracking performance in scenarios where the fundamental component of the signal becomes small or temporarily disappears.

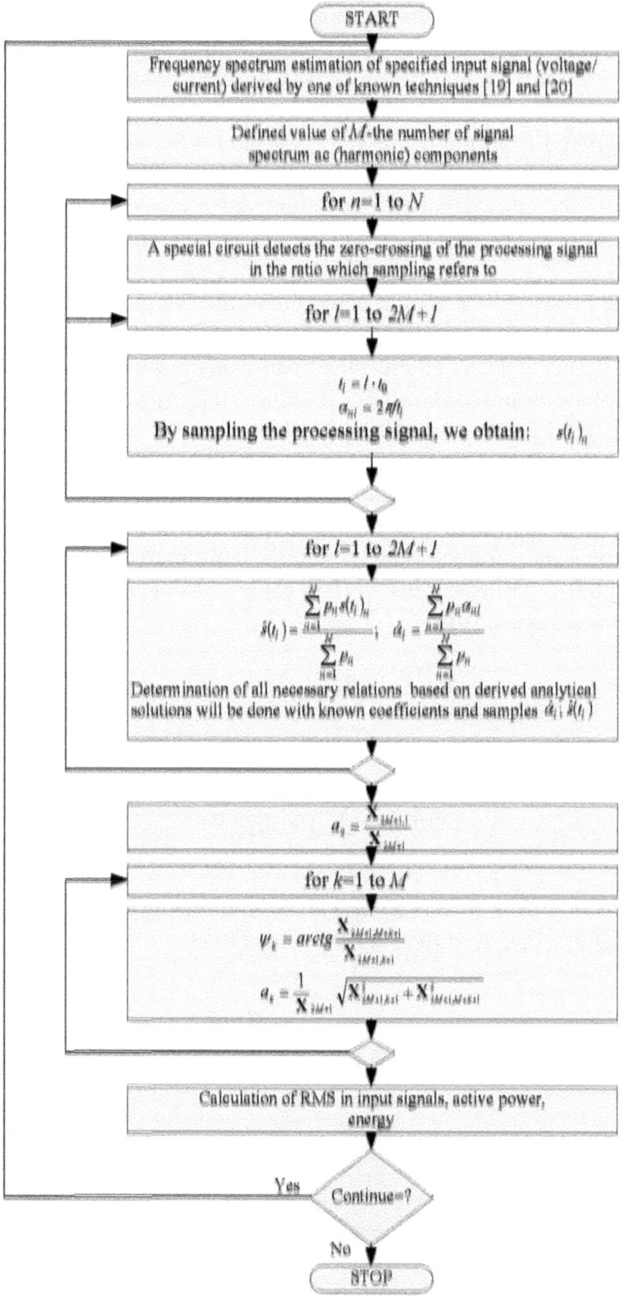

Fig. 3.1 Flow-chart of proposed reconstruction algorithm

In order to calculate the unknown parameters (amplitude and phase) of the processed periodic signals, the sampled values of the input analog signals $s(t_l)$ (as given in Eq. (3.2)) are required. These samples are obtained by sampling the signals at specific time instances, typically referenced to the detected zero-crossing moments. The values derived from the expressions depend on the measured frequency f, as the determinant elements are calculated based on the coefficient α_l derived from Eq. (3.5). However, other parameters in the derived system of equations are independent of the carrier signal frequency and the initial integration moment, as observed in [6].

Due to the presence of errors in determining the samples $s(t_l)$ and variables α_l, which are influenced by the carrier frequency f of the processed signal, it is essential to obtain the best estimates of these values based on the chosen criterion in practical applications of the proposed algorithm. To achieve this, a recalculation of the samples $s(t_l)$ and variables α_l is performed through N iterations (N is arbitrary) as outlined in the proposed algorithm. This process involves forming series of values, denoted as $s(t_l)_n$ and α_{nl} ($n = 1,\ldots,N$). The random errors Δ_n associated with the measurements are assumed to be unbiased, with an expected value $E(\Delta_n) = 0$. Additionally, the errors have the same variance $\mathrm{var}(\Delta_n) = \sigma^2$ and are mutually uncorrelated. Under these assumptions, the least squares (LS) estimator is employed to minimize the residual sum of squares [19]:

$$S\big(\hat{s}(t_l)\big) = \sum_{n=1}^{N} p_n\big(\hat{s}(t_l) - s(t_l)_n\big)^2 \tag{3.44}$$

$$S\big(\hat{\alpha}_l\big) = \sum_{n=1}^{N} p_n\big(\hat{\alpha}_l - \alpha_{nl}\big)^2 \tag{3.45}$$

where $p_n = m/\sigma^2, m > 0$ (m is arbitrary). By minimizing the function S, we obtain the LS estimators $\hat{s}(t_l)$, $\hat{\alpha}_l$ of the values $s(t_l)$ and α_l as:

$$\hat{s}(t_l) = \frac{\sum_{n=1}^{N} p_n s(t_l)_n}{\sum_{n=1}^{N} p_n} \tag{3.46}$$

$$\hat{\alpha}_l = \frac{\sum_{n=1}^{N} p_n \alpha_{nl}}{\sum_{n=1}^{N} p_n} \tag{3.47}$$

The choice of N in the estimation process depends on the desired processing speed, where higher values of N lead to more precise estimations. The LS method, known for its low computational complexity, has been widely applied in various engineering problems. In the specific case of the proposed algorithm, it does not require matrix inversion and is computationally less demanding compared to the methods described in [20]. When applied in simulations, the estimation of variables (Eqs. (3.46) and (3.47)) is not necessary, resulting in significant reduction of processing time.

To enhance the accuracy of the input signal samples, the proposed solution can be modified. In [21], it was demonstrated that implementing sampling and reconstruction with internal antialiasing filtering greatly improves the performance of digital receivers, leading to reduced errors in reconstruction.

It is important to note that there may exist frequency mismatches (FM) between the proposed signal frequencies and the actual frequencies in real applications. In [22, 23], a novel LMS-based Fourier analyzer was proposed to address FM and estimate discrete Fourier coefficients simultaneously. This analyzer effectively compensates for performance degradation caused by FM. However, in the LS procedure presented here, the estimation of samples $s(t_l)$ and variables αl is achieved by minimizing the S functions (3.44) and (3.45). Moreover, the derived analytical expressions are used to determine the unknown Fourier coefficients. Additionally, in each iteration of the described algorithm (Fig. 3.1), the sampling moment is referenced to the detected zero-crossing of the processed signal, and the basic frequency is simultaneously calculated. This reduces the dependence of determining the unknown parameters of the processed periodic signal on potential FM, in comparison to the algorithm analyzed in [22, 23]. The parameters analyzed in our paper exhibit less interdependence. However, if the signal-to-noise ratio is extremely low with significant FM, it is still possible to modify the estimation procedure as presented in [22, 23], without imposing additional requirements on the implementation.

Figure 3.2 depicts the impact of frequency estimation errors on the relative error of the system determinant for different harmonic contents of the input periodic signal. To enhance the algorithm's immunity, a more complex algorithm for detecting signal zero-crossing moments can be employed [24]. Uncertainty analysis in high-speed calibration systems is addressed in [25], which considers the effect of time base generator uncertainty (jitter) modeled as non-stationary additive noise. The paper also proposes a method for calculating uncertainty bounds around the reconstructed waveform based on a specified confidence level. The error resulting from the assumed non-idealities in the reconstruction model adheres to the boundaries specified in [25, 26]. A sensitivity function, typically formulated under the assumption of noise-free data, provides point-wise information about the reliability of the reconstructed signal prior to sampling. In [26], the minimum error bound for signal reconstruction is derived assuming noisy data.

In contrast to the procedure described in [6], the proposed algorithm is less sensitive to variations in the carrier signal frequency. The moments t_l at which the input signal is sampled can be completely random (asynchronous) and independent of the processed signal's frequency, as defined in Eq. (3.4). The time interval between consecutive samples primarily depends on the speed of the S/H (sample and hold) circuit and the AD (analog to digital) conversion circuit, which produce the numeric equivalent of the input signal samples. However, in practical implementations of sigma-delta ADCs, the moments t_l can be defined as $t_l = lt_{sample}$, where $t_{sample} = 1/$

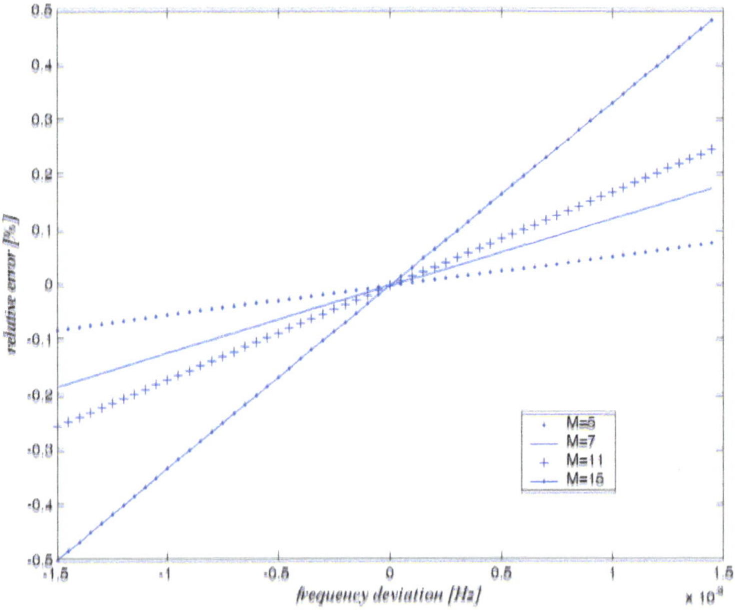

Fig. 3.2 Relative error in calculation of the system determinant as function of error in synchronization with frequency of fundamental harmonic of the input signal ($f = 50$ Hz, $t_l = 0.001$ s)

fs (fs denotes the sampling frequency). The start of sampling must be aligned with the detected zero crossing of the input signal.

3.2.1 *Numerical Complexity of Proposed Algorithm*

The derived relations involve multiplications and additions for realization (the derived relations require a total of $(2M+1)\left(4M^3+2M+1\right)$ multiplications and $(2M+1)\left(2^{2M}-\frac{1}{2}\binom{2M}{M}\right)$ additions). However, due to the method employed for determining the unknown parameters (as outlined in Fig. 3.1), the number of necessary operations is significantly reduced. Common factors in the formed products allow for their abbreviation during division operations. As a result, the total number of operations is reduced to $18M^2+12M+2$, indicating that the proposed algorithm requires $\frac{9}{2}(2M+1)^2$ flops.

The analytical expressions (3.16)–(3.43) are theoretically correct, while the error that arises during implementation is a result of the numerical procedures used for addition and multiplication. The values of potential errors are defined and analyzed in [27], where it is also demonstrated that the algorithms solving Vandermond-like systems (Eq. (3.3)) exhibit higher accuracy (although not more backward stable) compared to GEPP (requiring $\frac{2}{3}(2M+1)^3$ flops) or algorithms utilizing QR factorization (requiring $\frac{4}{3}(2M+1)^3$ flops).

It is well-known that the computational load of the iterative step involving FFTs remains unchanged regardless of the number of sampling points in some non-matrix implementations, but the convergence speed improves with a larger number of points available [28]. However, a larger number of sampling points leads to the presence of large matrices, especially when dealing with input signals with a wide spectrum. Consequently, the computational load of standard matrix methods (whether iterative or utilizing pseudo-inverse matrices) increases rapidly. While they may be efficient for situations with few sampling points, they tend to be slow when dealing with many points. In contrast, the proposed algorithm is non-iterative, resulting in faster computation. The required number of samples for the proposed algorithm to perform reconstruction in the observed system (3.3) is solely determined by the number of unknowns ($2M+1$ parameters). This characteristic makes our derived analytical solution computationally attractive for moderately-sized problems [27], and also viable for large-scale reconstruction problems. Additionally, the method is unaffected by spectral leakage, a common issue in reconstruction algorithms relying on FFT usage [28].

3.2.2 Computing Time

The proposed algorithm is well-suited for operation with sigma-delta ADC, offering high resolution and processing speed for input signals. This is a significant advantage when comparing its implementation in this context to the results presented in [7]. The time required to perform the necessary number of input signal samplings for reconstruction can be approximated as $(2M+1)t_{sample}$, representing the time needed for reconstruction in simulations. In practical applications, the reconstruction time needs to account for the additional time required to estimate the variables $s(t_l)$ and α_l (which depends on the value of N) and the time interval Δt needed for other recalculations according to the proposed algorithm (Fig. 3.1). Considering these factors, the reconstruction time can be defined as $N(2M+1)t_{sample} + \Delta t \approx N/f + \Delta t$, which also takes into account the synchronization with the input signal's zero crossing. The speed of the proposed algorithm is comparable to the algorithms analyzed in [29, 30].

In Ref. [31], the required processor time for signal reconstruction using matrix methods, as commonly implemented in program packages, is measured. The method proposed in this paper does not require special memorization of the transformation matrix or recalculation of the inversion matrix. This makes it more efficient to implement and not limited to sparse matrices.

Furthermore, this procedure can also be used for spectral analysis, allowing the determination of amplitude and phase values of signal harmonics based on the predicted system of equations. By following a systematic approach in conducting this procedure, it is possible to accurately establish the spectral content and optimize the proposed algorithm accordingly. In practical applications, the accuracy of signal reconstruction can be guaranteed in noisy environments by employing a powerful processor and appropriate filtering techniques.

3.3 Simulation and Experimental Results

The proposed algorithm was subjected to additional testing through simulation using the Matlab and SIMULINK software packages. In the presence of noise, the reconstruction process becomes an estimation task, where the reconstructed signal may vary due to the influence of the actual noise record. We specifically investigate the impact of noise and jitter on the estimation method. The presence of noise and jitter introduces errors in the detection of signal zero-crossing moments, leading to incorrect calculations of the derived relations. Additionally, errors arise in determining the values of the processed signal samples due to imprecisions in sampling time moments and the inherent errors of the real ADC.

Since the proposed reconstruction algorithm is based on standard hardware components used for sampling the input analog signal, we modeled the system as depicted in Fig. 3.3 for simulation purposes. The system included a sigma-delta ADC along with a sample-and-hold circuit for conversion. The frequency of the carrier signal was determined using a comparator (Schmitt-triggers) according to the method suggested in [24]. This setup allowed us to emulate the practical hardware implementation of the algorithm accurately.

The "Multi-harmonic input signal" block generates the input signal by combining multiple harmonic components. To account for the impact of noise and jitter on signal zero-crossing detection, a separate model is employed, which simulates their effects on the processed signal using the proposed algorithm. It is important to note that all Taylor expansions of jitter are biased [32]; therefore, incorporating jitter correctly

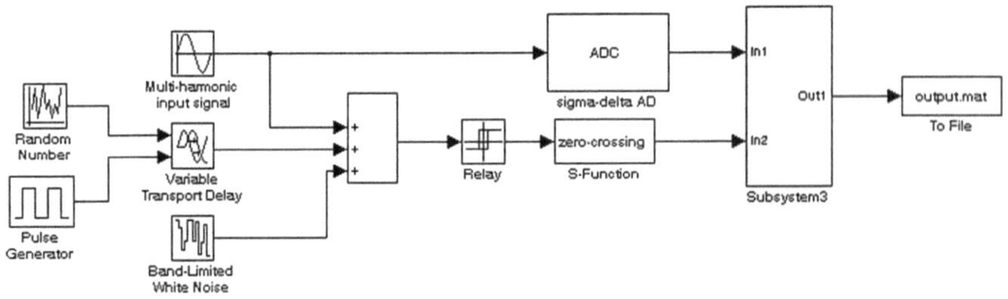

Fig. 3.3 SIMULINK model of the circuit for the realization of proposed reconstruction algorithm

involves introducing a variable delay in the signal path. The model for simulating jitter consists of a Pulse Generator and Random Number, whose outputs are fed into the circuit with Variable Transport Delay. Figure 3.3 illustrates the addition of white Gaussian noise to the complex periodic signal. In the original toolbox, all potential noise sources (primarily operational amplifiers and voltage references) were assumed to be white. The noise power in the "Band-Limited White Noise" block is represented by the Power Spectral Density (PSD), measured in V^2/Hz. By utilizing the Relay block depicted in Fig. 3.3, the signal generated in this manner is transformed into a series of right-angle impulses, allowing the measurement of the fundamental harmonic frequency within the "S-Function" block [24].

For the sampling and determination of the numeric equivalent of the input power signal samples, a sigma-delta ADC model described in [33] was utilized. During the simulation, the ADC had an effective resolution of 24 bits and a sampling rate of $f_S = 1$ kHz. The output Power Spectral Density (PSD) for the ideal signal, as well as for the signal affected by thermal noise and clock jitter, ranged from -100 to -170 dB for the signal-to-noise distortion ratio (SNDR) between 85 and 96 dB.

The ADC model incorporates a quantizer circuit, and the quantization error is a significant concern. This is due to the sophisticated nature of the proposed reconstruction algorithm, where certain operations, such as determinant calculation, are highly sensitive to quantization errors and can amplify them significantly.

Once the input signal samples are obtained using the aforementioned ADC model, they are introduced into the "Subsystem3" block along with the measured frequency of the input signal. This enables the determination of the unknown amplitudes and phases of the input periodic signal based on derived relationships. The simulation uses the parameter values specified in Table 3.3 for the input signal.

A signal comprising the first 7 harmonics with a fundamental frequency of $f =$ 50 Hz was employed. Table 3.3 provides the amplitude and phase values for the signal. In this simulation, the presence of superimposed noise and jitter results in a relative error in fundamental frequency detection of 0.0001%. The accuracy of the proposed algorithm aligns with the performance achieved in processing signals of similar nature, as reported in [30, 34]. The errors in amplitude and phase detection

Table 3.3 Simulation results of signal reconstruction by the proposed algorithm

Harmonic number	Amplitude (Vpp)	Phase (rad)	Proposed reconstruction algorithm	
			Amp. error (%)	Phase error (%)
1	1	π	0.0016	0.0021
2	0.73	$\pi/3$	0.0023	0.0018
3	0.64	0	0.0021	0.0022
4	0.55	$\pi/6$	0.0017	0.0019
5	0.32	$\pi/4$	0.0018	0.0024
6	0.27	$\pi/12$	0.0023	0.0017
7	0.14	0	0.0024	0.0023

primarily stem from inaccuracies in measuring input signal samples and determining the values of derived equations.

The investigation focused on examining the impact of noise on signal characteristics by evaluating the frequency and magnitude estimation of noisy signals. To conduct the study, a sinusoidal test signal with a frequency of 50 Hz was selected as the input, and additive white centered Gaussian noise was introduced. The intention was to achieve a predetermined Signal-to-Noise Ratio (SNR), represented by SNR $= 20\log\left(A/\sqrt{2}\sigma\right)$, where A denotes the magnitude of the signal's fundamental harmonics and σ represents the standard deviation of the noise. Figure 3.4 illustrates the maximum errors observed in frequency and harmonic magnitude estimates for input signals of 30, 50, and 70 Hz with SNRs ranging from 40 to 70 dB. It is noteworthy that, in practical scenarios, the SNR of voltage signals obtained from power systems typically falls within the range of 50 to 70 dB. Consequently, the proposed technique exhibits minimal errors at this level of noise, as depicted in Fig. 3.4 [35].

To explore the statistical properties of the proposed estimator, computer simulations were conducted using noisy samples. The generation of noisy samples involved adding white noise samples to the processed signal samples. When estimating deterministic parameters, a widely employed benchmark for assessing accuracy is the Cramer-Rao lower bound (CRLB). The CRLB, calculated as the inverse of the Fisher information, represents a lower limit on the mean squared error (MSE). Figures 3.5 and 3.6 present the MSE values for the amplitudes and frequency, respectively, based on 105 simulation runs. The results unequivocally demonstrate that the proposed estimation scheme converges asymptotically to the CRLB, indicating its effectiveness [35].

3.3.1 Experimental Results

To validate the simulation findings, experiments were performed using a laboratory prototype of a "star" connected induction machine. The control structure of the machine comprises six thyristors, with each pair of back-to-back connected thyristors linked between the phase source voltage and the corresponding phase of the induction machine in the "star" configuration. The specific parameters of the induction machine include a wound rotor type with two pole pairs, with $R_S = 5.2\Omega, R_r = 14.63\Omega, L_S = L_r = 0.055H, L_m = 1.3H$. The firing angle α is defined as the angle between the zero crossings of the phase voltage and the commencement of conduction for the corresponding thyristor.

The experimental setup comprises voltage and current transducers, a connector block, a data acquisition card, and a processing computer (PC). In this setup, an NI ELVIS/PCI-6251 Bundle acquisition card from National Instruments is employed to capture the current and voltage signals from each phase. These signals are then processed using the proposed algorithm on the PC. The measurement hardware

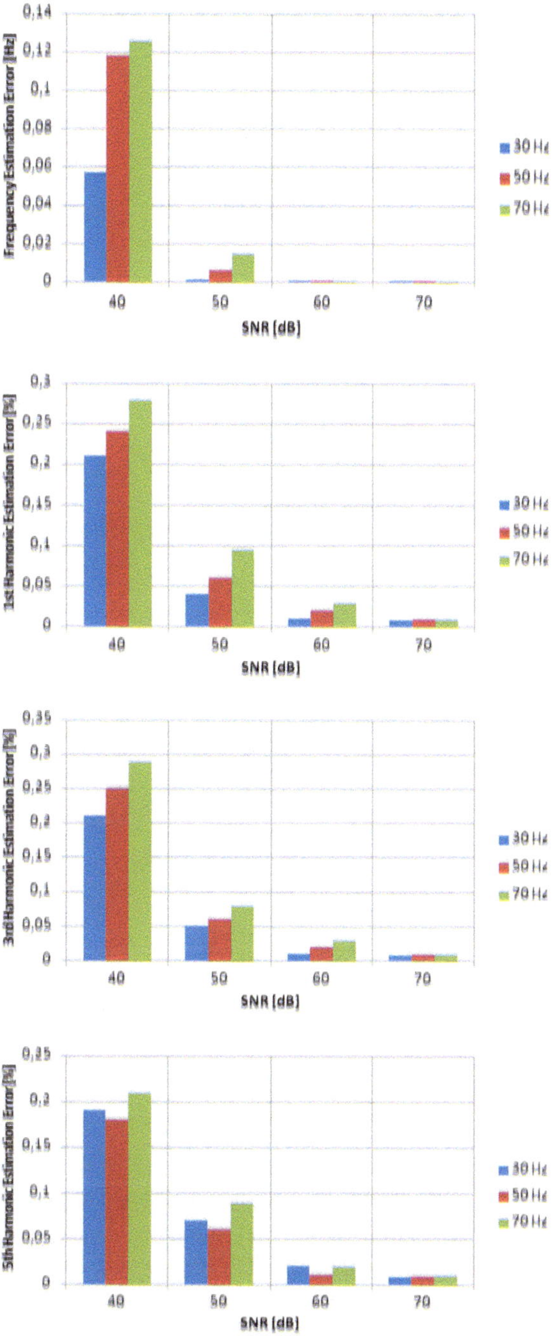

Fig. 3.4 Maximum estimations errors for noisy input signals

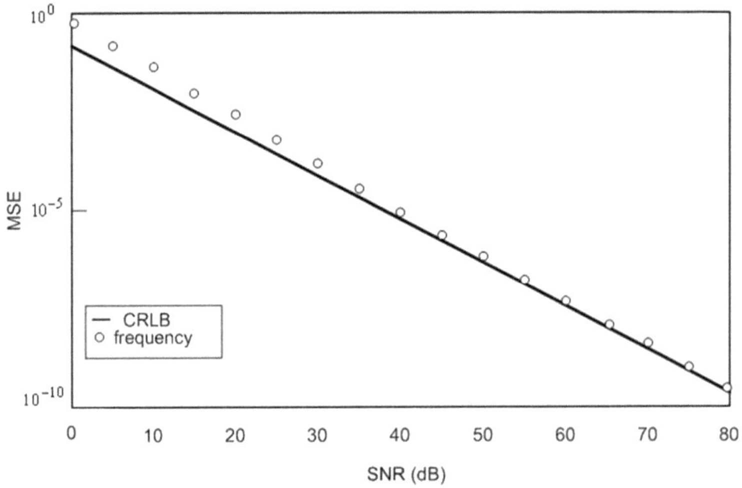

Fig. 3.5 MSE of the frequency as a function of SNR

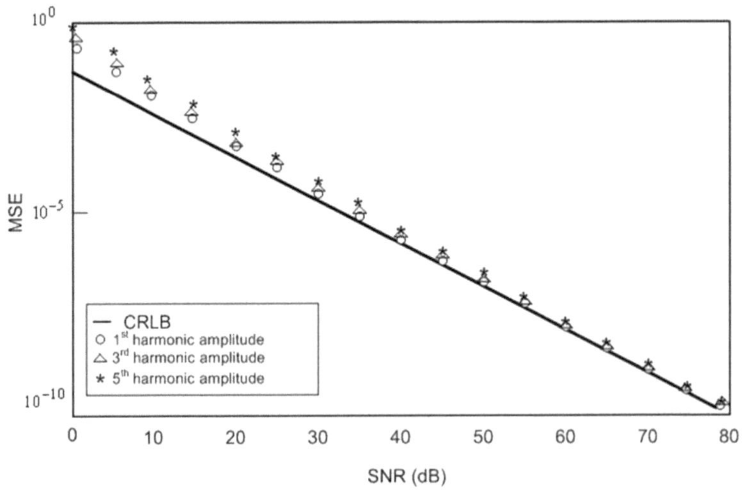

Fig. 3.6 MSE of the three harmonic amplitudes as a function of SNR

includes LEM HTR current and LEM CV3 voltage transducers. The sampling rate of the DAQ card is set to 250 kS/s.

For the synchronous speed of the machine ($n = 1500$ rpm) and the "star" connection configuration, the parameters of current and voltage harmonics in phase 'a' are recorded using NI LabVIEW software, as shown in Table 3.4. Additionally, the signals with this harmonic content are processed using the proposed algorithm to estimate signal parameters. The amplitude and phase detection differences between the data acquisition card and the proposed reconstruction algorithm are also presented in Table 3.4. The analysis demonstrates the high accuracy of the proposed algorithm in processing periodic signals under real-world conditions.

The proposed algorithm, described in [12], is a novel approach that offers complexity reduction for estimating Fourier coefficients. The derived analytical expression enables fast calculations of essential signal parameters, such as phase and amplitude, with minimal numerical errors. A standard digital signal processor (DSP) and a real sigma-delta ADC are sufficient to meet the hardware requirements. Moreover, the suggested concept can be utilized as a standalone algorithm for spectral analysis of processed signals. By identifying the parameters of AC signals, it becomes possible to determine relevant values in electric utilities, including energy, power, and RMS values. The measurement uncertainty is influenced by errors in

Table 3.4 Comparison of experimental results by the proposed reconstruction algorithm and results obtained with acquisition card NI ELVIS/PCI-6251 Bundle

Parameters	Acquisition card ($\alpha = 110°$)		Proposed reconstruction algorithm ($\alpha = 110°$)	
	Measured values	Phase (deg)	Amplitude differe. (%)	Phase differe. (%)
I_{ph_a1} (1st harmonic) (A)	0.63	57.87	0.0021	0.0024
I_{ph_a3} (3rd harmonic) (A)	0	0	0.0019	0.0025
I_{ph_a5} (5th harmonic) (A)	0.13	−21.77	0.0026	0.0022
I_{ph_a7} (7th harmonic) (A)	0.04	−80.21	0.0018	0.0028
I_{ph_a11} (11th harmonic) (A)	0.1	154.85	0.0018	0.0021
I_{ph_a13} (13th harmonic) (A)	0.114	25.47	0.0013	0.0023
I_{ph_a17} (17th harmonic) (A)	0.02	−12.63	0.0029	0.0017
I_{ph_a19} (19th harmonic) (A)	0.029	−56.48	0.0023	0.0027
V_{out_a1} (1th harmonic) (V)	271.94	0	0.0018	0
V_{out_a3} (3rd harmonic) (V)	33.62	122.23	0.0028	0.0023
V_{out_a5} (5th harmonic) (V)	23.12	−42.37	0.0017	0.0026
V_{out_a7} (7th harmonic) (V)	10.52	74.68	0.0017	0.0023

synchronizing with the fundamental frequency of the processing signal (due to non-stationary nature of jitter-related noise and white Gaussian noise) as well as errors in determining the values of the processed signal samples. Simulation results confirm that the proposed algorithm achieves satisfactory precision in reconstructing periodic signals within real-world environments.

3.4 Reconstruction of Analogue Multi-harmonic Signals, from a Number of Integrated Values of Input Signals

The study conducted in [36] investigates the application of integrative sampling for reconstructing periodic signals. Traditional sampling involves a sample and hold circuit followed by analog-to-digital conversion using an AD converter. However, [36] explores the possibility of eliminating the separate sample and hold circuit, which is an original approach. This research lays the foundation for a novel algorithm designed for processing slowly varying AC signals. The integration technique described in the paper is employed to process the original analog input signal, which is often corrupted by noise. An important advantage of the proposed method is that it filters both random and periodic disturbances through integrative sampling.

In simulation scenarios, where integrative samples are assumed to be error-free, the presented algorithm can be used for reconstructing periodic band-limited signals without further modifications. However, in real-world environments, integrative samples are subject to measurement errors. These errors arise from imprecision introduced by voltage and current transducers, the conversion process, and the digital circuits involved in signal processing. Consequently, the sampled values obtained in practice may not precisely represent the signal at sampling points but rather serve as averages of the signal in their vicinity. In such cases, modifications to the suggested algorithm are necessary to determine the optimal signal estimate based on the chosen criterion.

The approach outlined in [36] relies on utilizing the values derived from integral processing of the continuous input signal within defined time intervals. This integral processing is performed repeatedly to enable the reconstruction of the multi-harmonic signal under consideration. This approach leads to the formulation of a specific system of linear equations, similar to Eq. (3.3).

By integrating the signal (3.1) and averaging it over the integration interval (similar to Eq. (3.7) [37]), and by establishing a system of equations with the same form to determine the $2M + 1$ unknowns (including amplitudes, phases of the M harmonics, and the average value of the signal), the following equation is derived:

$$x(t_l) = \frac{1}{\Delta t}\int\limits_{t_l-\frac{\Delta t}{2}}^{t_l+\frac{\Delta t}{2}}\left[a_0 + \sum_{k=1}^{M} a_k \sin(k2\pi f_0\tau + \psi_k)\right]d\tau$$

$$= a_0 - \frac{1}{2\pi f\Delta t}\sum_{k=1}^{M}\frac{a_k}{k}\left[\cos\left(k2\pi f_0\left(t_l + \frac{\Delta t}{2}\right) + \psi_k\right) - \cos\left(k2\pi f_0\left(t_l - \frac{\Delta t}{2}\right) + \psi_k\right)\right]$$

$$= a_0 + \frac{1}{\pi f\Delta t}\sum_{k=1}^{M}\frac{a_k}{k}\sin(2k\pi f_0 t_l + \psi_k)\sin(k\pi f_0\Delta t) \tag{3.48}$$

In the given equation, where $l = 1, 2, \ldots, 2M + 1$, the time interval $(t_l + \Delta t/2) - (t_l - \Delta t/2) = \Delta t = const.$ (for each l) corresponds to the interval over which the input analog signal is integrated. The time instant $t_l - \Delta t/2$ is defined as the starting moment for the integration process. With each subsequent integration, this moment needs to be adjusted accordingly. By averaging the input signal within the integration interval, we simplify the derived formulas. This type of processing aligns with ADC transducers that provide average values over the integration time, such as dual-slope AD converters. The value $x(t_l)$ represents the average value of the integral and is determined based on a reference value utilized in the conversion process or the functioning of the circuit responsible for the integration. This relationship can be concisely represented as:

$$a_0 + \sum_{k=1}^{M} A_k a_k\left(\sin\alpha_{k,l}\cos\psi_k + \cos\alpha_{k,l}\sin\psi_k\right) = x(t_l) \tag{3.49}$$

where:

$$\frac{1}{k\pi f_0\Delta t}\sin[k\pi f_0\Delta t] = A_k;\ (k = 1, 2, \ldots, M);$$
$$2k\pi f_0 t_l = \alpha_{k,l};\ (l = 1, 2, \ldots, 2M + 1) \tag{3.50}$$

The variables A_k and $\alpha_{k,l}$ correspond to the outcomes of signal integration in the physical device, as described by Eq. (3.1). The subsequent initiation time for the integration process can be defined as follows:

$$t_l = t_{l-1} + \Delta t + t_{delay} \tag{3.51}$$

In the above Eq. (3.51), where t_{delay} denotes the processing delay, the value of Δt, which is a constant, is arbitrary. The determinant of the system comprising $2M + 1$ unknown parameters can be expressed as follows:

$$\mathbf{X}_{system} = \begin{vmatrix} 1 & A_1 \sin\alpha_{1,1} & \ldots & A_M \sin\alpha_{M,1} & A_1 \cos\alpha_{1,1} & \ldots & A_M \cos\alpha_{M,1} \\ 1 & A_1 \sin\alpha_{1,2} & \ldots & A_M \sin\alpha_{M,2} & A_1 \cos\alpha_{1,2} & \ldots & A_M \cos\alpha_{M,2} \\ \cdot & \cdot & \cdot\ldots & & & \ldots & \\ 1 & A_1 \sin\alpha_{1,2M+1} & \ldots & A_M \sin\alpha_{M,2M+1} & A_1 \cos\alpha_{1,2M+1} & \ldots & A_M \cos\alpha_{M,2M+1} \end{vmatrix}$$
$$= (A_1 \cdot A_2 \cdot \ldots \cdot A_{M-1} \cdot A_M)^2 \mathbf{X}_{2M+1} \tag{3.52}$$

where:

$$\mathbf{X}_{2M+1} = \begin{vmatrix} 1 & \sin\alpha_{1,1} & \sin\alpha_{2,1} & \ldots & \sin\alpha_{M,1} & \cos\alpha_{1,1} & \cos\alpha_{2,1} & \ldots & \cos\alpha_{M,1} \\ 1 & \sin\alpha_{1,2} & \sin\alpha_{2,2} & \ldots & \sin\alpha_{M,2} & \cos\alpha_{1,2} & \cos\alpha_{2,2} & \ldots & \cos\alpha_{M,2} \\ \cdot & \cdot & \cdot & & & & & & \\ 1 & \sin\alpha_{1,2M+1} & \sin\alpha_{2,2M+1} & \ldots & \sin\alpha_{M,2M+1} & \cos\alpha_{1,2M+1} & \cos\alpha_{2,2M+1} & \ldots & \cos\alpha_{M,2M+1} \end{vmatrix}$$
$$= \begin{vmatrix} 1 & \sin\alpha_1 & \sin 2\alpha_1 & \ldots & \sin M\alpha_1 & \cos\alpha_1 & \cos 2\alpha_1 & \ldots & \cos M\alpha_1 \\ 1 & \sin\alpha_2 & \sin 2\alpha_2 & \ldots & \sin M\alpha_2 & \cos\alpha_2 & \cos 2\alpha_2 & \ldots & \cos M\alpha_2 \\ \cdot & \cdot & \cdot & & & & & & \\ 1 & \sin\alpha_{2M+1} & \sin 2\alpha_{2M+1} & \ldots & \sin M\alpha_{2M+1} & \cos\alpha_{2M+1} & \cos 2\alpha_{2M+1} & \ldots & \cos M\alpha_{2M+1} \end{vmatrix} \tag{3.53}$$

Here:

$$\alpha_{k,l} = k\alpha_l = 2k\pi f_0 t_l \tag{3.54}$$

It is evident that Eq. (3.53) is identical to Eq. (3.6). The unknown parameters of the signal, namely amplitude and phase, can be obtained by dividing the solution expression of the corresponding cofactors by the analytical solution expression of the system determinant.

$$\begin{aligned} \psi_k &= arctg \frac{\mathbf{X}_{2M+1,M+k+1}}{\mathbf{X}_{2M+1,k+1}} \\ a_k &= \frac{1}{A_k \mathbf{X}_{2M+1}} \sqrt{\mathbf{X}^2_{2M+1,k+1} + \mathbf{X}^2_{2M+1,M+k+1}} \end{aligned} \tag{3.55}$$

In practical applications of the proposed algorithm, it is essential to obtain the most accurate estimates of the integrative samples $x(t_l)$, variables A_k and α_l, considering the inherent error caused by their dependency on the carrier frequency f_0 of the processed signal (as shown in Figs. 3.4, 3.5, and 3.6). To achieve this, a recalculation process can be performed through N iterations (where N is arbitrary), resulting in series $x(t_l)$, A_k and α_l. The measurement errors Δ_i are assumed to be unbiased, with an expected value $E(\Delta_i) = 0$, having the same variance $\text{var}(\Delta_i) = \sigma^2$, and being uncorrelated. Under these assumptions, a weighted average approach can be employed to reduce random errors in determining the observed values. Weighted averages are commonly used when dealing with uncorrelated measurements of varying accuracy. The average values $\hat{x}(t_l)$, $\hat{A}_k$, $\hat{\alpha}_l$ are computed for all $l = 1, 2,\ldots, 2M + 1$ and $k = 1, 2,\ldots, M$ using the following expressions:

$$\hat{x}(t_l) = \frac{\sum_{i=1}^{n_x} w_{xi} x(t_l)_i}{\sum_{i=1}^{n_x} w_{xi}}; \hat{A}_k = \frac{\sum_{n=1}^{n_A} w_{Ai} A_{ki}}{\sum_{i=1}^{n_A} w_{Ai}}; \hat{\alpha}_l = \frac{\sum_{n=1}^{n_\phi} w_{\phi i} \alpha_{li}}{\sum_{i=1}^{n_\phi} w_{\phi i}};$$

$$\sum_{i=1}^{n_x} w_{xi} = \sum_{i=1}^{n_A} w_{Ai} = \sum_{i=1}^{n_\phi} w_{\phi i} = N \tag{3.56}$$

Here, $w_{xi}, w_{Ai}, w_{\phi i}$, represents non-negative weights associated with the series $x(t_l)_i$, A_{ki}, α_{li}. The variable n_x, n_A, n_ϕ, indicates the number of distinct values in the aforementioned series obtained through N iterations. The choice of N depends on the desired processing speed, with higher values of N resulting in more accurate estimations. Notably, the estimation procedure in this specific case does not necessitate matrix inversion and is computationally less demanding.

Due to the presence of noise, the reconstruction of the signal becomes an estimation task, resulting in variations in the reconstructed signal depending on the actual noise encountered. The estimated input signal in the measurement system is subject to two types of errors: systematic and stochastic. In our proposed system for signal reconstruction, we have successfully eliminated the sample-and-hold circuit as a potential source of systematic errors. Furthermore, we have conducted an investigation into the impact of noise and jitter on the measurement method. The presence of noise and jitter can lead to false detection of signal zero-crossing moments, which subsequently results in incorrect calculations of derived relationships. Additionally, errors arise when determining the values of the processed signal samples. These errors stem from imprecise determination of the sampling time as well as the inherent inaccuracies introduced by the real analog-to-digital converter (ADC).

To analyze the errors, we performed a comprehensive error analysis using both Matlab programming and SIMULINK simulation. Figure 3.7 illustrates the influence of frequency determination errors in the carrier signal on the relative error in determining the integration interval and the A_k coefficients (Eq. (3.50)). Figures 3.8 and 3.9 present another dependency, depicting how the relative error in determining the

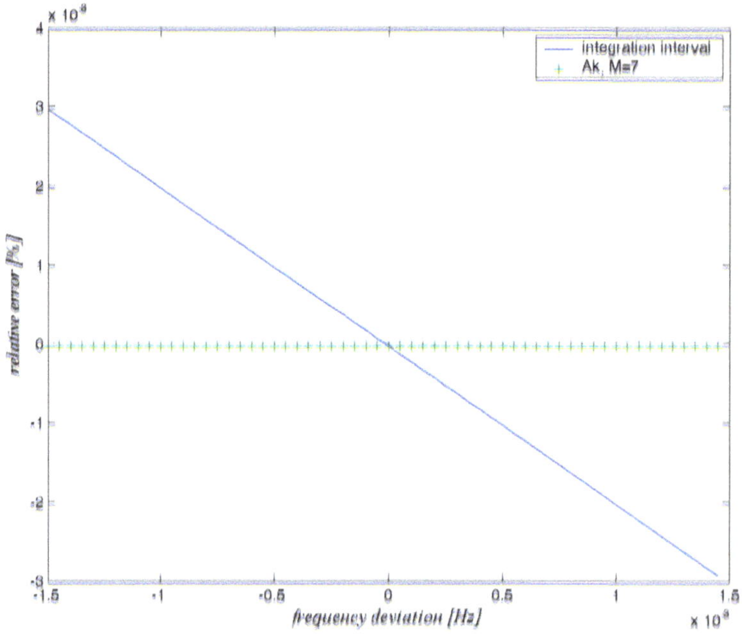

Fig. 3.7 Relative error in determination of integration time and variables A_k as function of error in synchronization with frequency of fundamental harmonic of the input signal

value of the input signal integral and the value of the system determinant vary with the carrier signal frequency determination errors for different harmonic contents of the input periodic signal. Notably, the error in determining the integration interval is independent of the harmonic content. Our analyses have shown that the best approach for determining the starting moment of integration is by aligning it with the zero-crossing of the input signal ($t_1 - \Delta t/2 = 0$).

Quantization error poses a significant challenge as the proposed reconstruction algorithm is of a sophisticated nature. Certain operations, such as determinant calculation, are particularly prone to amplifying quantization errors due to their inherently ill-conditioned nature. However, in this context, the quantization error is considerably minimized due to the described concept of measuring (integration) of the input signal and the hardware implementation as depicted in Fig. 3.10.

In the proposed algorithm, the highest harmonic component's order M in the processed signal spectrum must be known or adopted in advance, even if it exceeds the expected (real) value.

To recalculate the unknown parameters (amplitude and phase) of the processed periodic signals, the results of integrating the input analog signals $x(t_l)$ (as given in Eqs. (3.49) and (3.50)) are required. The integral samples of the input signal are obtained by integrating within precisely defined time intervals of the signal under reconstruction. The starting moment $t_l - \Delta t/2$ of the integration process is referenced to the detected zero crossing of the input analog signal. It is possible to set $t_1 - \Delta t/2$

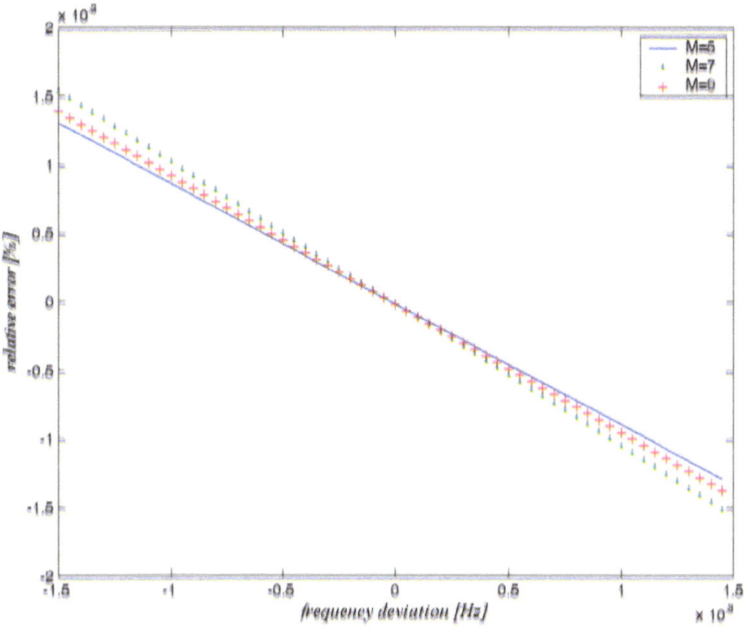

Fig. 3.8 Relative error in integral calculation of input signal as function of error in synchronization with frequency of fundamental harmonic of the input signal

as 0 (the starting moment of the integration process in the first iteration). Since each subsequent integration follows immediately after the previous one, the total integration time in each iteration is practically determined by $t_{\text{int.},l} = \Delta t + t_{delay}$. In the suggested algorithm, the carrier signal frequency is recalculated in each iteration to account for any possible changes. This recalculation is incorporated into the process of determining the new integration interval for the input analog signal, the new variables A_k, α_l and the value of the input analog signal's integral $x(t_l)$. This approach minimizes the error in the reconstruction process resulting from variations in the frequency of the processed signal. The integration interval Δt is taken as a constant value to enable the proposed method. By setting its value as $\Delta t = 1/(f_0(2M + 1))$, the integration process covers the entire period of the complex input signal. This ensures that all possible signal changes within its period are accounted for during processing, and it reduces the error in calculating the variable A_k. The proposed procedure relies on the harmonic frequency f_0, as all relevant parameters calculated in the algorithm are dependent on f_0. However, the issue of synchronization is less pronounced compared to standard techniques of synchronized sampling, as this involves the actual reconstruction of a signal to determine its fundamental parameters (phase and amplitude).

It is important to note that the values of the determinants and co-determinants are influenced by the measured frequency of the fundamental harmonic of the processed signal, as well as the chosen constant and the initial moment $t_1 - \Delta t/2$ from which the

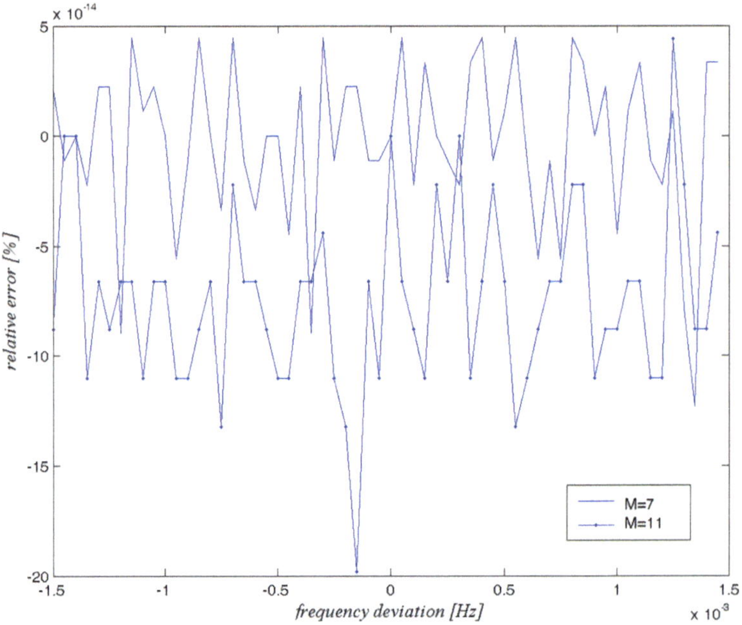

Fig. 3.9 Relative error in calculation of the system determinant as function of error in synchronization with frequency of fundamental harmonic of the input signal

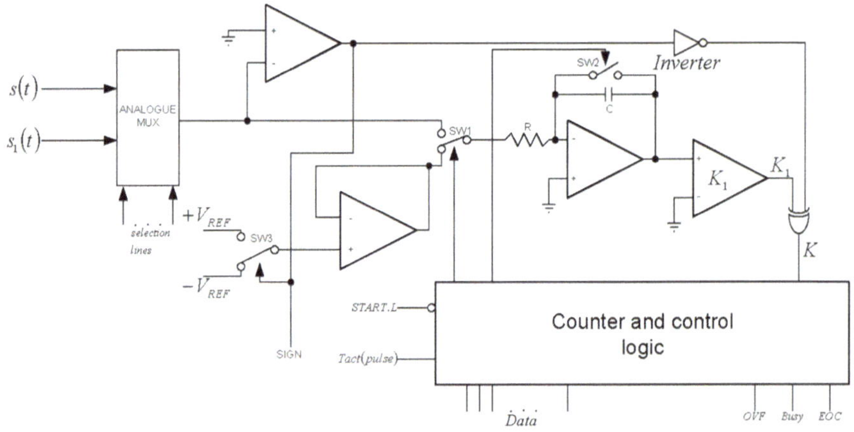

Fig. 3.10 Block scheme of the digital circuit for the realization of proposed method of processing

integration process begins. This is because the determinant elements are calculated based on the coefficient α_l, as defined in Eq. (3.54).

The proposed algorithm can be effectively implemented in conjunction with sigma-delta ADC, offering advantages in terms of high resolution and processing speed of input signals. This distinction should be taken into consideration when comparing its implementation to the results presented in [37]. To realize the integration circuit, a separate integrator circuit can be employed, incorporating a reset option to mitigate the impact of time constants on the precision of integrating the input analog signal. This resetting operation is performed internally in dual-slope and sigma-delta ADCs. Subsequently, the counter generates a digital equivalent of the integral obtained in this manner (Fig. 3.10). Based on this, it is feasible to realize the circuit in a straightforward manner as an integrated circuit (IC).

The time required to perform the necessary number of integrations for reconstructing the input signal is defined as $(2M + 1)(\Delta t + t_{delay})$ and directly depends on the system's operating frequency. It can be practically limited to the period of the processed signal, which approximates the reconstruction time in simulations. In practical applications of the proposed algorithm, the determined reconstruction time should be extended by the estimation time $t_{estimation}$ required to estimate the variables $x(t_l)$, A_l and α_l. This estimation time is directly related to the value of N, although it can be restricted to a maximum interval of 1 s. The time interval t_{delay} depends on the speed at which the capacitor in the integrator circuit (Fig. 3.10) approaches zero and is influenced by the switching speed of the switches and the analog multiplexer. This time can be reduced by leveraging the processing principles used in sigma-delta ADCs. The speed of the proposed algorithm is comparable to the algorithms analyzed in [29, 30].

In [31], the required processor time for signal reconstruction using the matrix method is measured. This method, as implemented in many software packages, necessitates the storage of transformation matrices and the recalculation of inversion matrices. In contrast, the approach proposed in this paper is more efficient, as it does not require special memorization of transformation matrices and does not necessitate the recalculation of inversion matrices. It is also adaptable to non-sparse matrices and is more straightforward to implement in hardware. The proposed algorithm can be practically implemented on various platforms.

Furthermore, this procedure can be applied to spectral analysis, enabling the determination of amplitude and phase values of signal harmonics based on the anticipated system of equations. By incrementally conducting this procedure, the precise spectral content can be established, followed by optimization of the proposed algorithm to align it with the actual signal characteristics. To ensure accurate signal reconstruction in noisy environments, the use of a powerful processor with appropriate filtering techniques is recommended.

3.4.1 Possible Hardware Realization of Proposed Method of Processing and Simulation Results

Figure 3.10 depicts the block diagram of the digital circuit used to implement the proposed processing method.

Figure 3.10 illustrates that there is no dedicated sample-and-hold circuit in the digital circuit implementation. Instead, the analog signal is directly fed into the integrating circuit through an analog multiplexer, which determines the processing type. The input periodic signals $s(t)$ and $s_1(t)$, such as voltage and current signals, undergo adaptation to the measurement range of the processing components.

Within a precisely defined interval, the counter counts up to a specific value, which is stored in the accompanying logic (the DSP controlling the entire system in Fig. 3.10, where the proposed processing algorithm is implemented). Subsequently, the counter is reset. The digital equivalent of the calculated value is determined using the counter and a reference voltage signal, following the same principle employed in dual-slope or sigma-delta ADCs. After the integrator performs the recalculation according to Eq. (3.48) during the second time interval, dependent on the counter's speed (counter's clock signal), a numerical representation of the calculated integral of the analog input signal is established using a known reference value. This value is then transferred to the controlling processor via the "DATA" lines (Fig. 3.7). At the end of the second interval, the comparator at the integrator output identifies the completion of the conversion process when the total voltage at the integrator output reaches zero. This generates a signal that resets the counter and resets the capacitor in the integrating circuit, preparing the suggested system for a new integration process. Special selection lines control the operation of the analog multiplexer at the input, determining the type of signal (voltage or current) being processed. The proposed solution can be modified to reduce the error in determining the integral of the input signal, allowing for nonideal but feasible signal reconstruction even without a dedicated addition to perform the best estimation of the observed signal's integrative samples $x(t_l)$.

The value $x(t_l)$ (as described in Eqs. (3.49) and (3.50)) represents the integral of the input analog signal. It is worth noting that the sign of V_{REF} depends on the sign of $x(t_l)$. If $x(t_l)$ is positive, V_{REF} must be negative, and vice versa. The "SIGN" line in Fig. 3.10 indicates the detected polarity of the input signal, allowing the polarity of the reference voltage to be selected through a separate switch. The sign of the expression $x(t_l)$ can be verified programmatically, adjusting the sign of V_{REF} accordingly. This can be avoided by conducting operations using absolute values. To assess the effectiveness of the proposed hardware solution, a simulation was conducted using the SIMULINK software package (Fig. 3.11).

Figure 3.11 illustrates the inclusion of various noise components, including white Gaussian noise, thermal noise, $1/f$ noise, and jitter, into the complex periodic signal. In the original toolbox, it was assumed that all possible noise sources, primarily from operational amplifiers and voltage references, were white. The noise power in the "Band-Limited White Noise" block represents the Power Spectral Density (PSD), measured in V^2/Hz. Since noise power is additive, the PSD can be considered as the sum of flicker ($1/f$) noise and thermal noise terms, associated with sampling switches and the intrinsic noise of the operational amplifier. To accurately analyze the sampling noise and op-amp's thermal and flicker noise at low frequencies, separate blocks were employed in Fig. 3.11. Here, k, T, and C represent Boltzmann's constant, temperature in Kelvin, and the sampling capacitor, respectively [33]. V_n denotes

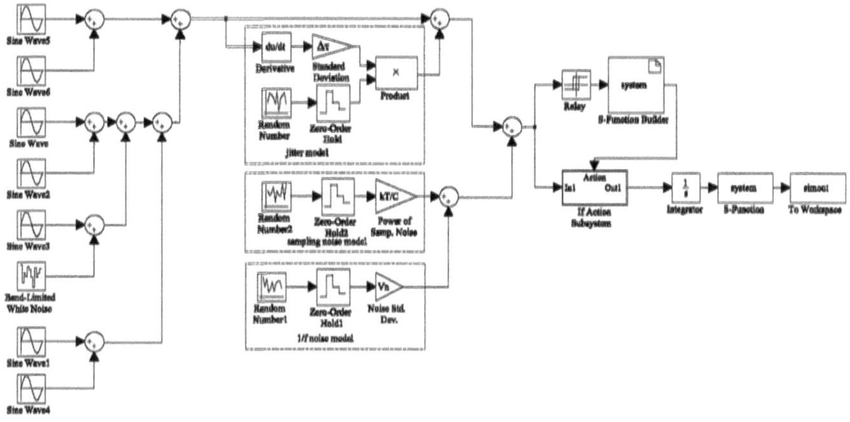

Fig. 3.11 SIMULINK model of proposed hardware realization

the input-referred thermal noise of the op-amp, encompassing flicker noise, wide-band thermal noise, and DC offset contributions. The total noise power ${V^2}_n$ can be evaluated through transistor level simulation, with V_n assumed to be 30 μVrms and the sampling capacitance set as $C = 2$ pF. Jitter effects were simulated using a model depicted in Fig. 3.8, assuming uncorrelated Gaussian random process with a standard deviation $\Delta\tau$. The jitter model was extensively described and analyzed in [32, 38], demonstrating its frequency dependency on the input sinusoidal frequency. During the simulation, a jitter value of 1 ns (standard deviation $\Delta\tau$) was utilized [33, 39]. This simulation approach resulted in output PSD values ranging from -100 to -170 dB for the Signal-to-Noise Distortion Ratio (SNDR) between 56 and 96 dB.

Using the Relay block depicted in Fig. 3.11, the frequency of the fundamental harmonic was measured in the signal generated through this process, generating a series of samples for input data in the "S-Function Builder" block. Within this block, the integration interval was defined based on the proposed algorithm, subsequently generating the signal action (enable signal) in the "If Action Subsystem" block. The enable signal determines the time interval during which integration of the input signal takes place. Once the integrals of the input signals are determined through this procedure, they are passed to the "S-Function" block to establish the values of the unknown amplitudes and phases of the input periodic signal based on derived relations. The implementation of the already developed MATLAB/SIMULINK model and code on the dSPACE 1104 platform is straightforward. During the simulation, the parameters of the input signal align with the values provided in Table 3.5.

A signal comprising the first 7 harmonics, with a fundamental frequency $f_0 =$ 50 Hz, was utilized for the simulation. Table 3.5 provides the amplitude and phase values of the signal. The superimposed noise and jitter, as simulated in this study, resulted in a relative error of 0.0001% in the detection of the fundamental frequency. In the time domain, the relative error between the original signal and its reconstruction was found to be 0.00256%. The inaccuracies observed in the amplitude and

Table 3.5 Comparison of simulation results of signal reconstruction by the proposed algorithm and FFT

Harmonic number	Amplitude (Vpp)	Phase (rad)	Proposed reconstruction algorithm		FFT (sampling rate = 25 kHz; data length = 25,000; time period = 1 s)	
			Amp. error (%)	Phase error (%)	Amp. error (%)	Phase error (%)
1	1	π	0.0019	0.0024	0.296	0.322
2	0.73	π/3	0.0026	0.0022	0.035	0.038
3	0.64	0	0.0024	0.0025	0.875	0.843
4	0.55	π/6	0	0	0	0
5	0.32	π/4	0	0	0	0
6	0.27	π/12	0	0	0	0
7	0.14	0	0	0	0	0

phase detection primarily stem from errors in measuring the integrative samples and determining the values of the derived equations.

The proposed algorithm [36] holds significant promise for signal reconstruction and highly accurate measurements of periodic signals. Our implementations have enhanced the computational efficiency of this algorithm for moderately sized problems and even made it feasible for large-scale reconstruction tasks. The approach is based on integrative sampling of input analog signals, and the derived analytical expression enables real-time calculations of fundamental signal parameters, such as phase and amplitude. The required hardware resources can be accommodated by a standard digital signal processor (DSP). By eliminating the need for a separate sample-and-hold circuit, a potential source of system errors has been eliminated. Moreover, the proposed concept can be employed as a standalone algorithm for spectral analysis of processed signals. Utilizing the identified parameters of the AC signals, important values in electric utilities, such as energy, power, and RMS, can be determined. An error analysis of the signal reconstruction process has been conducted, taking into account the uncertainties arising from synchronization with the fundamental frequency of the processing signal (attributed to the non-stationary nature of jitter-related noise and white Gaussian noise) as well as errors in determining the integral samples of the processed signal. The results demonstrate that the proposed algorithm maintains high accuracy in reconstructing periodic signals in real-world scenarios.

3.5 Estimation of Amplitude and Phase of Analog Multiharmonic Signal Based on the Differential Irregular Samples

The aforementioned approach [40, 41] is centered around the utilization of values obtained through differential processing of the continuous input signal, within well-defined time intervals. This repeated processing enables the reconstruction of the multi-harmonic single-tone analog input signal, which is the focal point of the operation. As a result, a specific form of a linear equations system is derived, which can be efficiently solved using the derived analytical expressions. The algorithm for estimating the amplitude and phase of the complex AC signal requires $O((2M)^2)$ floating-point operations, whereas the direct solution of the derived equations necessitates $O((2M)^3)$ floating-point operations. Thus, the proposed method significantly enhances computational efficiency compared to standard reconstruction algorithms, while maintaining a lower numerical error.

By performing differentiation on the signal (3.1) and establishing a system of equations with a similar structure, we can determine the $2M$ unknowns, which correspond to the amplitudes and phases of the M harmonics. The system of equations is as follows:

$$\frac{d(s(t))}{dt} = \frac{d\left(\sum\limits_{k=1}^{M} a_k \sin(k2\pi f t + \psi_k)\right)}{dt}\bigg|_{t=t_l} = x(t_l) \Rightarrow$$

$$x(t_l) = \sum_{k=1}^{M} a_k k2\pi f \cos(k2\pi f t_l + \psi_k) =$$

$$= \sum_{k=1}^{M} a_k k2\pi f[\cos(k2\pi f t_l)\cos(\psi_k) - \sin(k2\pi f t_l)\sin(\psi_k)] \tag{3.57}$$

For each $l = 1, 2,\ldots, 2M$, the equation system consists of differentiations performed on the input analog signal at specific time moments t_l. The t_l values are determined by the speed of the differentiation circuit, and they need to be adjusted for each subsequent differentiation. The $x(t_l)$ represents the differentiated value of the processed signal. In a concise representation, the relation can be expressed as:

$$\sum_{k=1}^{M} A_k a_k\left(\cos\alpha_{k,l}\cos\psi_k - \sin\alpha_{k,l}\sin\psi_k\right) = x(t_l) \tag{3.58}$$

where:

$$k2\pi f = A_k; \quad 2k\pi f t_l = \alpha_{k,l}; \quad (k = 1, 2, \ldots, M); (l = 1, 2, \ldots, 2M) \tag{3.59}$$

The variables A_k and $\alpha_{k,l}$ are obtained as a result of differentiating the signal in the real device, as defined by Eq. (3.57). These variables are determined by the sampling moment and the frequency of the corresponding harmonic in the input periodic signal. The system determinant for the system of $2M$ unknown parameters can be expressed as:

$$\mathbf{X}_{system} = \begin{vmatrix} A_1 \cos\alpha_{1,1} & \ldots & A_M \cos\alpha_{M,1} & -A_1 \sin\alpha_{1,1} & \ldots & -A_M \sin\alpha_{M,1} \\ A_1 \cos\alpha_{1,2} & \ldots & A_M \cos\alpha_{M,2} & -A_1 \sin\alpha_{1,2} & \ldots & -A_M \sin\alpha_{M,2} \\ . & .\ldots & . & . & .\ldots & . \\ A_1 \cos\alpha_{1,2M} & \ldots & A_M \cos\alpha_{M,2M} & -A_1 \sin\alpha_{1,2M+1} & \ldots & -A_M \sin\alpha_{M,2M} \end{vmatrix}$$
$$= (-1)^M (A_1 \cdot A_2 \cdot \ldots \cdot A_{M-1} \cdot A_M)^2 \mathbf{X}_{2M} \tag{3.60}$$

where:

$$\mathbf{X}_{2M} = \begin{vmatrix} \cos\alpha_{1,1} & \cos\alpha_{2,1} & \ldots & \cos\alpha_{M,1} & \sin\alpha_{1,1} & \sin\alpha_{2,1} & \ldots & \sin\alpha_{M,1} \\ \cos\alpha_{1,2} & \cos\alpha_{2,2} & \ldots & \cos\alpha_{M,2} & \sin\alpha_{1,2} & \sin\alpha_{2,2} & \ldots & \sin\alpha_{M,2} \\ . & . & . & . & . & . & . & . \\ \cos\alpha_{1,2M} & \cos\alpha_{2,2M} & \ldots & \cos\alpha_{M,2M} & \sin\alpha_{1,2M+1} & \sin\alpha_{2,2M} & \ldots & \sin\alpha_{M,2M} \end{vmatrix}$$
$$= \begin{vmatrix} \cos\varphi_1 & \cos 2\varphi_1 & \ldots & \cos M\varphi_1 & \sin\varphi_1 & \sin 2\varphi_1 & \ldots & \sin M\varphi_1 \\ \cos\varphi_2 & \cos 2\varphi_2 & \ldots & \cos M\varphi_2 & \sin\varphi_2 & \sin 2\varphi_2 & \ldots & \sin M\varphi_2 \\ . & . & . & . & . & . & . & . \\ \cos\varphi_{2M} & \cos 2\varphi_{2M} & \ldots & \cos M\varphi_{2M} & \sin\varphi_{2M} & \sin 2\varphi_{2M} & \ldots & \sin M\varphi_{2M} \end{vmatrix} \tag{3.61}$$

Here:

$$\varphi_l = \frac{\alpha_{k,l}}{k} = 2\pi f t_l \tag{3.62}$$

In the system of equations derived from the proposed signal processing concept (Eq. (3.58)), the trigonometric values for $\varphi_1, \varphi_2, \varphi_3, \ldots, \varphi_{2M}$ (as defined in Eq. (3.62)) need to be used instead of the x variables found in [36]. The given determinant (Eq. (3.61)) can be transformed by utilizing *Euler's* formulas as follows:

$$
\begin{aligned}
\mathbf{X}_{2M} &= \left| \cos\varphi_1 \;\; \cos 2\varphi_1 \;\; \ldots \;\; \cos M\varphi_1 \;\; \sin\varphi_1 \;\; \sin 2\varphi_1 \;\; \ldots \;\; \sin M\varphi_1 \right| \\
&= \frac{e^{-M\frac{\pi}{2}i}}{2^{2M}} \left| e^{\varphi_1 i} + e^{-\varphi_1 i} \;\; e^{2\varphi_1 i} + e^{-2\varphi_1 i} \;\; \ldots \;\; e^{M\varphi_1 i} + e^{-M\varphi_1 i} \;\; e^{\varphi_1 i} - e^{-\varphi_1 i} \;\; e^{2\varphi_1 i} - e^{-2\varphi_1 i} \;\; \ldots \;\; e^{M\varphi_1 i} - e^{-M\varphi_1 i} \right| \\
&= \frac{e^{-M\frac{\pi}{2}i}}{2^{M}} \left| e^{\varphi_1 i} \;\; e^{2\varphi_1 i} \;\; \ldots \;\; e^{M\varphi_1 i} \;\; e^{\varphi_1 i} - e^{-\varphi_1 i} \;\; e^{2\varphi_1 i} - e^{-2\varphi_1 i} \;\; \ldots \;\; e^{M\varphi_1 i} - e^{-M\varphi_1 i} \right| \\
&= \frac{(-1)^M}{2^M} e^{-M\frac{\pi}{2}i} \left| e^{\varphi_1 i} \;\; e^{2\varphi_1 i} \;\; \ldots \;\; e^{M\varphi_1 i} \;\; e^{-\varphi_1 i} \;\; e^{-2\varphi_1 i} \;\; \ldots \;\; e^{-M\varphi_1 i} \right| \\
&= \frac{(-1)^{\frac{M(M-1)}{2}}}{2^M} e^{-M\frac{\pi}{2}i} \left| e^{-M\varphi_1 i} \;\; \ldots \;\; e^{-\varphi_1 i} \;\; e^{\varphi_1 i} \;\; \ldots \;\; e^{M\varphi_1 i} \right| \\
&= \frac{(-1)^{\frac{M(M-1)}{2}}}{2^M} e^{-M\frac{\pi}{2}i} e^{-M(\varphi_1+\varphi_2+\ldots+\varphi_{2M})i} \left| 1 \;\; e^{\varphi_1 i} \;\; \ldots \;\; e^{(M-1)\varphi_1 i} \;\; e^{(M+1)\varphi_1 i} \;\; \ldots \;\; e^{2M\varphi_1 i} \right| \\
&= \frac{(-1)^{\frac{M(M-1)}{2}}}{2^M} e^{-M\frac{\pi}{2}i} e^{-M(\varphi_1+\varphi_2+\ldots+\varphi_{2M})i} \Delta_{2M+1,M+1}\Big|_{x_j=e^{\varphi_j i}}
\end{aligned}
$$

$$
\Delta_{2M+1,M+1}\Big|_{x_j=e^{\varphi_j i}} = \Delta_{2M+1,M+1}^{(2M+1)}\Big|_{x_j=e^{\varphi_j i}} = \frac{\Delta_{2M+1}}{\prod_{k=1}^{2M}(x_{2M+1}-x_k)}(x_1x_2\ldots x_{2M})\sum \frac{1}{(x_1x_2\ldots x_{2M})_M}\Big|_{x_j=e^{\varphi_j i}}
$$

$$
\Delta_{2M+1}\Big|_{x_j=e^{\varphi_j i}} = 2^{M(2M+1)} e^{M(2M+1)\frac{\pi}{2}i} e^{M(\varphi_1+\varphi_2+\ldots+\varphi_{2M})i} \prod_{j=k+1}^{2M+1}\prod_{k=1}^{2M} \sin\frac{\varphi_j-\varphi_k}{2}
$$

$$
\prod_{k=1}^{2M}(x_{2M+1}-x_k)\Big|_{x_j=e^{\varphi_j i}} = (-1)^M 2^{2M} e^{M\varphi_{2M+1} i} e^{\frac{1}{2}(\varphi_1+\varphi_2+\ldots+\varphi_{2M})i} \prod_{k=1}^{2M} \sin\frac{\varphi_{2M+1}-\varphi_k}{2} \Rightarrow
$$

$$
\mathbf{X}_{2M} = (-1)^{\frac{M(M-1)}{2}} 2^{2M(M-1)} \frac{\prod_{j=k+1}^{2M+1}\prod_{k=1}^{2M}\sin\frac{\varphi_j-\varphi_k}{2}}{\prod_{k=1}^{2M}\sin\frac{\varphi_{2M+1}-\varphi_k}{2}} \sum \cos\left\{\frac{\varphi_1+\varphi_2+\ldots+\varphi_{2M}}{2} - (\varphi_1+\varphi_2+\ldots+\varphi_{2M})_M\right\}
$$

$$
\mathbf{X}_{2M} = (-1)^{\frac{M(M-1)}{2}} 2^{2M(M-1)} \prod_{j=k+1}^{2M}\prod_{k=1}^{2M-1} \sin\frac{\varphi_j-\varphi_k}{2} \sum \cos\left\{\frac{\varphi_1+\varphi_2+\ldots+\varphi_{2M}}{2} - (\varphi_1+\varphi_2+\ldots+\varphi_{2M})_M\right\} \tag{3.63}
$$

The co-determinants necessary to solve the given system of equations (3.60) are as follows:

$$
\mathbf{X}_{2M,1} = \begin{vmatrix} x(t_1) & \cos 2\varphi_1 & \ldots & \cos M\varphi_1 & \sin\varphi_1 & \sin 2\varphi_1 & \ldots & \sin M\varphi_1 \\ x(t_2) & \cos 2\varphi_2 & \ldots & \cos M\varphi_2 & \sin\varphi_2 & \sin 2\varphi_2 & \ldots & \sin M\varphi_2 \\ \cdot & \cdot & \cdot & \cdot & \cdot & \cdot & \cdot & \cdot \\ x(t_{2M}) & \cos 2\varphi_{2M} & \ldots & \cos M\varphi_{2M} & \sin\varphi_{2M} & \sin 2\varphi_{2M} & \ldots & \sin M\varphi_{2M} \end{vmatrix} \tag{3.64}
$$

$$
\mathbf{X}_{2M,2} = \begin{vmatrix} \cos\varphi_1 & x(t_1) & \ldots & \cos M\varphi_1 & \sin\varphi_1 & \sin 2\varphi_1 & \ldots & \sin M\varphi_1 \\ \cos\varphi_2 & x(t_2) & \ldots & \cos M\varphi_2 & \sin\varphi_2 & \sin 2\varphi_2 & \ldots & \sin M\varphi_2 \\ \cdot & \cdot & \cdot & \cdot & \cdot & \cdot & \cdot & \cdot \\ \cos\varphi_{2M} & x(t_{2M}) & \ldots & \cos M\varphi_{2M} & \sin\varphi_{2M} & \sin 2\varphi_{2M} & \ldots & \sin M\varphi_{2M} \end{vmatrix} \tag{3.65}
$$

and so on. Based on the aforementioned derivation, the co-determinants can be expressed as follows:

$$\mathbf{X}_{2M,1} = x(t_1)\mathbf{X}_1^1 + x(t_2)\mathbf{X}_2^1 + \ldots + x(t_{2M})\mathbf{X}_{2M}^1 \tag{3.66}$$

The co-determinants, denoted as $\mathbf{X}_1^1, \mathbf{X}_2^1, \ldots, \mathbf{X}_{2M}^1$, are obtained by eliminating the corresponding row and the first column from the co-determinant $\mathbf{X}_{2M,1}$ after expansion. The second co-determinant (or co-factor) is derived from the expansion of $\mathbf{X}_{2M}$ along such a column. For this purpose, $\mathbf{X}_p^q$ needs to be determined as co-factors of $\mathbf{X}_{2\mathrm{M}}$, therefore:

$$\mathbf{X}_p^q = (-1)^{p+q}\mathbf{F}_p^q \tag{3.67}$$

where $\mathbf{F}_p^q$ is obtained from X_{2M} overturning p row and q column. We know that:

$$\Delta_{2M+1,M+1} = \Delta_{2M+1,M+1}^{(2M+1)} \tag{3.68}$$

$$\mathbf{X}_{2M} = \frac{(-1)^{\frac{M(M-1)}{2}}}{2^M} e^{-M\frac{\pi}{2}i} e^{-M(\varphi_1+\ldots+\varphi_{2M})i} \Delta_{2M+1,M+1}\left(x_t = e^{\varphi_t i}\right) \tag{3.69}$$

Hence, it follows that $\mathbf{X}_p^q$ is co-factor of matrix, whose determinant is $\Delta_{2M+1,M+1}$.

$$\Delta_{2M+1,M+1} = \left(\prod_{j=k+1}^{2M}\prod_{k=1}^{2M-1}\left(x_j - x_k\right)\right)(x_1x_2\ldots x_{2M})\sum\frac{1}{(x_1x_2\ldots x_{2M})_M} \tag{3.70}$$

For $1 \le p \le 2M \wedge 1 \le q \le M$:

$$
\begin{aligned}
\mathbf{F}_p^q &= \left|\cos\varphi_1\ \ldots\ \cos(q-1)\varphi_1\ \cos(q+1)\varphi_1\ \ldots\ \cos M\varphi_1\ \sin\varphi_1\ \ldots\ \sin M\varphi_1\right|_p = \\
&= \frac{e^{-M\frac{\pi}{2}i}}{2^{2M-1}}\left|e^{\varphi_1 i}+e^{-\varphi_1 i}\ \ldots\ e^{(q-1)\varphi_1 i}+e^{-(q-1)\varphi_1 i}\ e^{(q+1)\varphi_1 i}+e^{-(q+1)\varphi_1 i}\ \ldots\ e^{M\varphi_1 i}+e^{-M\varphi_1 i}\ e^{\varphi_1 i}-e^{-\varphi_1 i}\ \ldots\ e^{M\varphi_1 i}-e^{-M\varphi_1 i}\right|_p = \\
&= \frac{e^{-M\frac{\pi}{2}i}}{2^{M}}\left|e^{-\varphi_1 i}\ \ldots\ e^{-(q-1)\varphi_1 i}\ e^{-(q+1)\varphi_1 i}\ \ldots\ e^{-M\varphi_1 i}\ e^{\varphi_1 i}\ \ldots\ e^{(q-1)\varphi_1 i}\ e^{q\varphi_1 i}-e^{-q\varphi_1 i}\ e^{(q+1)\varphi_1 i}\ \ldots\ e^{M\varphi_1 i}\right|_p = \\
&= \frac{e^{-M\frac{\pi}{2}i}}{2^{M}}\left\{\begin{array}{l}\left|e^{-\varphi_1 i}\ \ldots\ e^{-(q-1)\varphi_1 i}\ e^{-(q+1)\varphi_1 i}\ \ldots\ e^{-M\varphi_1 i}\ e^{\varphi_1 i}\ \ldots\ e^{M\varphi_1 i}\right|_p+ \\ +(-1)^M\left|e^{-\varphi_1 i}\ \ldots\ e^{-M\varphi_1 i}\ e^{\varphi_1 i}\ \ldots\ e^{(q-1)\varphi_1 i}\ e^{(q+1)\varphi_1 i}\ \ldots\ e^{M\varphi_1 i}\right|_p\end{array}\right\} = \\
&= \frac{(-1)^{\frac{M(M+1)}{2}}}{2^M}e^{-M\frac{\pi}{2}i}\left\{\begin{array}{l}\left|e^{-M\varphi_1 i}\ \ldots\ e^{-\varphi_1 i}\ e^{\varphi_1 i}\ \ldots\ e^{(q-1)\varphi_1 i}\ e^{(q+1)\varphi_1 i}\ \ldots\ e^{M\varphi_1 i}\right|_p- \\ -\left|e^{-M\varphi_1 i}\ \ldots\ e^{-(q+1)\varphi_1 i}\ e^{-(q-1)\varphi_1 i}\ \ldots\ e^{-\varphi_1 i}\ e^{\varphi_1 i}\ \ldots\ e^{M\varphi_1 i}\right|_p\end{array}\right\} = \\
&= \frac{(-1)^{\frac{M(M+1)}{2}}}{2^M}e^{-M\frac{\pi}{2}i}e^{-M\left(\varphi_1+\ldots+\varphi_{p-1}+\varphi_{p+1}+\ldots+\varphi_{2M}\right)i}\left\{\begin{array}{l}\left|1\ \ldots\ e^{(M-1)\varphi_1 i}\ e^{(M+1)\varphi_1 i}\ \ldots\ e^{(M+q-1)\varphi_1 i}\ e^{(M+q+1)\varphi_1 i}\ \ldots\ e^{2M\varphi_1 i}\right|_p- \\ -\left|1\ \ldots\ e^{(M-q-1)\varphi_1 i}\ e^{(M-q+1)\varphi_1 i}\ \ldots\ e^{(M-1)\varphi_1 i}\ e^{(M+1)\varphi_1 i}\ \ldots\ e^{2M\varphi_1 i}\right|_p\end{array}\right\} = \\
&= \frac{(-1)^{\frac{M(M+1)}{2}}}{2^M}e^{-M\frac{\pi}{2}i}e^{-M\left(\varphi_1+\ldots+\varphi_{p-1}+\varphi_{p+1}+\ldots+\varphi_{2M}\right)i}\left(\Delta_{2M+1,M+1}^{(p,M+q)}-\Delta_{2M+1,M+1}^{(p,M-q+1)}\right)_{x_j=e^{\varphi_j i}}
\end{aligned}
\tag{3.71}
$$

It follows that:

$$\mathbf{F}_p^q = \frac{(-1)^{\frac{M(M+1)}{2}}}{2^M} e^{-M\frac{\pi}{2}i} e^{-M(\phi_1+\ldots+\phi_{p-1}+\phi_{p+1}+\ldots+\phi_{2M})i}$$
$$\left(\Delta_{2M+1,M+1}^{(p,M+q)} - \Delta_{2M+1,M+1}^{(p,M-q+1)}\right)_{x_j=e^{\phi_j i}}$$
$$for\ 1 \le p \le 2M \wedge 1 \le q \le M \tag{3.72}$$

where $\Delta_{2M+1,M+1}^{(r,s)}$ is the determinant obtained from $\Delta_{2M+1,M+1}$ after the r row and s column have been eliminated.

$$\Delta_{2M+1,M+1} = \begin{vmatrix} 1 & x_1 & \ldots x_1^{M-1} & x_1^{M+1} & \ldots x_1^{2M} \\ \ldots & \ldots & \ldots & \ldots & \ldots \\ 1 & x_{2M} & \ldots x_{2M}^{M-1} & x_{2M}^{M+1} & \ldots x_{2M}^{2M} \end{vmatrix} \Rightarrow$$
$$\Delta_{2M+1,M+1} = \left(\prod_{j=k+1}^{2M} \prod_{k=1}^{2M-1} (x_j - x_k)\right)(x_1x_2\ldots x_{2M}) \sum \frac{1}{(x_1x_2\ldots x_{2M})_M} \tag{3.73}$$

When we determinate $\Delta_{2M+1,M+1}^{(r,s)}$, we must eliminate r row and s column from $\Delta_{2M+1,M+1}$, and if $\Delta_{2M+1,M+1}$ is developed by r row, what we obtain is that $D_{r,s} = (-1)^{r+s}\Delta_{2M+1,M+1}^{(r,s)}$ is the coefficient in $x_r^{s-1} (1 \le s \le M)$, i.e. the coefficient beside x_r^s if $M + 1 \le s \le 2M$.

$$\Delta_{2M+1,M+1} = (-1)^r(x_r - x_1)\ldots(x_r - x_{r-1})(x_r - x_{r+1})\ldots(x_r - x_{2M})\cdot$$
$$\cdot\left(\prod_{\substack{1\le j\le 2M \\ k,j\ne r}} \prod_{\substack{1\le k\le 2M \\ k<j}} (x_j - x_k)\right)(x_1x_2\ldots x_{2M}) \sum \frac{1}{(x_1x_2\ldots x_{2M})_M} \tag{3.74}$$

Here is:

$$(x_1x_2 \ldots x_{2M}) \sum \frac{1}{(x_1x_2\ldots x_{2M})_M} = (x_1\ldots x_{r-1}x_{r+1}\ldots x_{2M})\cdot$$
$$\cdot\, x_r \cdot \left(\sum \frac{1}{(x_1\ldots x_{r-1}x_{r+1}\ldots x_{2M})_M} + \frac{1}{x_r}\sum \frac{1}{(x_1\ldots x_{r-1}x_{r+1}\ldots x_{2M})_{M-1}}\right) \Rightarrow$$
$$\Delta_{2M+1,M+1}^{(r,s)} = \left(\prod_{\substack{1\le j\le 2M \\ k,j\ne r}} \prod_{\substack{1\le k\le 2M \\ k<j}} (x_j - x_k)\right)(x_1\ldots x_{r-1}x_{r+1}\ldots x_{2M})$$

$$\begin{Bmatrix} \sum (x_1 \ldots x_{r-1}x_{r+1} \ldots x_{2M})_{2M-s} \cdot \sum \dfrac{1}{(x_1 \ldots x_{r-1}x_{r+1} \ldots x_{2M})_{M-1}} - \\ - \sum (x_1 \ldots x_{r-1}x_{r+1} \ldots x_{2M})_{2M-s+1} \cdot \sum \dfrac{1}{(x_1 \ldots x_{r-1}x_{r+1} \ldots x_{2M})_{M}} \end{Bmatrix}$$
$$for\ 1 \le s \le M \tag{3.75}$$

If we introduce the following symbols:

$$\begin{aligned} V_t &= \sum (x_1 \ldots x_{r-1}x_{r+1} \ldots x_{2M})_t \Big|_{x_j = e^{\varphi_j i}} \\ \overline{V}_t &= \sum \frac{1}{(x_1 \ldots x_{r-1}x_{r+1} \ldots x_{2M})_t} \Big|_{x_j = e^{\varphi_j i}} \end{aligned} \tag{3.76}$$

It follows that:

$$\begin{aligned} &\Delta_{2M+1,M+1}^{(p,M-q+1)}(x_t = e^{\phi_t i}) = \left(\prod_{\substack{1 \le j \le 2M \\ k,j \ne p}} \prod_{\substack{1 \le k \le 2M \\ k < j}} (x_j - x_k) \right) (x_1 \ldots x_{p-1}x_{p+1} \ldots x_{2M}) \\ &\left(V_{M+q-1}\overline{V}_{M-1} - V_{M+q}\overline{V}_M \right)\Big|_{x_t = e^{\phi_t i}} \\ &for\ 1 \le q \le M-1,\ while\ for \\ &q = M \Rightarrow V_{2M} = 0;\ V_{2M-1} = (x_1 \ldots x_{p-1}x_{p+1} \ldots x_{2M}) \end{aligned} \tag{3.77}$$

We can write that:

$$\begin{aligned} &\Delta_{2M+1,M+1}^{(p,M+q)}(x_t = e^{\phi_t i}) = -\left(\prod_{\substack{1 \le j \le 2M \\ k,j \ne p}} \prod_{\substack{1 \le k \le 2M \\ k < j}} (x_j - x_k) \right) (x_1 \ldots x_{p-1}x_{p+1} \ldots x_{2M}) \\ &\left(V_{M-q-1}\overline{V}_{M-1} - V_{M-q}\overline{V}_M \right)\Big|_{x_t = e^{\phi_t i}} \\ &for\ 1 \le q \le M-2,\ while\ for\ q = M-1 \Rightarrow V_0 = 1 \\ &and\ for\ q = M \Rightarrow V_{-1} = 0;\ V_0 = 1 \end{aligned} \tag{3.78}$$

From this, it follows that:

$$\begin{aligned} \mathbf{F}_p^q &= \frac{(-1)^{\frac{M(M+1)}{2}}}{2^M} e^{-M\frac{\pi}{2}i} e^{-M(\phi_1 + \ldots + \phi_{p-1} + \phi_{p+1} + \ldots + \phi_{2M})i} \left(\prod_{\substack{1 \le j \le 2M \\ k,j \ne p}} \prod_{\substack{1 \le k \le 2M \\ k < j}} (x_j - x_k) \right) \cdot \\ &\cdot (x_1 \ldots x_{p-1}x_{p+1} \ldots x_{2M}) \begin{Bmatrix} (V_{M+q}\overline{V}_M - V_{M+q-1}\overline{V}_{M-1}) + \\ + (V_{M-q}\overline{V}_M - V_{M-q-1}\overline{V}_{M-1}) \end{Bmatrix} \Big|_{x_t = e^{\phi_t i}} \end{aligned} \tag{3.79}$$

$$\left(\prod_{\substack{1\le j\le 2M\\ k,j\ne p}}\prod_{\substack{1\le k\le 2M\\ k\prec j}}\left(x_j-x_k\right)\right)\Bigg|_{x_t=e^{\varphi_t i}}=(-1)^p\frac{\prod_{j=k+1}^{2M}\prod_{k=1}^{2M-1}\left(e^{\varphi_j i}-e^{\varphi_k i}\right)}{\prod_{\substack{1\le k\le 2M\\ k\ne p}}\left(e^{\varphi_p i}-e^{\varphi_k i}\right)} \tag{3.80}$$

We can write that:

$$\prod_{j=k+1}^{2M}\prod_{k=1}^{2M-1}\left(e^{\varphi_j i}-e^{\varphi_k i}\right)=(-1)^M 2^{M(2M-1)}e^{-M\frac{\pi}{2}i}e^{\frac{2M-1}{2}(\varphi_1+\ldots+\varphi_{2M})i}\prod_{j=k+1}^{2M}\prod_{k=1}^{2M-1}\sin\frac{\varphi_j-\varphi_k}{2}$$

$$\prod_{\substack{1\le k\le 2M\\ k\ne p}}\left(e^{\varphi_p i}-e^{\varphi_k i}\right)=2^{2M-1}e^{\frac{2M-1}{2}\pi i}e^{\frac{2M-1}{2}\varphi_p i}e^{\frac{1}{2}\left(\varphi_1+\ldots\varphi_{p-1}+\varphi_{p+1}+\ldots+\varphi_{2M}\right)i}\prod_{\substack{1\le k\le 2M\\ k\ne p}}\sin\frac{\varphi_p-\varphi_k}{2}\Rightarrow$$

$$\left(\prod_{\substack{1\le j\le 2M\\ k,j\ne p}}\prod_{\substack{1\le k\le 2M\\ k\prec j}}\left(x_j-x_k\right)\right)\Bigg|_{x_t=e^{\varphi_t i}}=(-1)^p e^{-(M-1)\frac{\pi}{2}i}2^{2M^2-3M+1}$$

$$e^{(M-1)\left(\varphi_1+\ldots\varphi_{p-1}+\varphi_{p+1}+\ldots+\varphi_{2M}\right)i}\frac{\prod_{j=k+1}^{2M}\prod_{k=1}^{2M-1}\sin\frac{\varphi_j-\varphi_k}{2}}{\prod_{\substack{1\le k\le 2M\\ k\ne p}}\sin\frac{\varphi_p-\varphi_k}{2}}\Rightarrow$$

$$\mathbf{F}_p^q=(-1)^p(-1)^{\frac{M(M-1)}{2}}i\cdot 2^{2M^2-4M+1}\frac{\prod_{j=k+1}^{2M}\prod_{k=1}^{2M-1}\sin\frac{\varphi_j-\varphi_k}{2}}{\prod_{\substack{1\le k\le 2M\\ k\ne p}}\sin\frac{\varphi_p-\varphi_k}{2}}\cdot$$
$$\cdot\left\{\left(V_{M+q}\overline{V}_M-V_{M+q-1}\overline{V}_{M-1}\right)+\left(V_{M-q}\overline{V}_M-V_{M-q-1}\overline{V}_{M-1}\right)\right\} \tag{3.81}$$

It follows that:

$$\frac{\mathbf{X}_p^q}{\mathbf{X}_{2M}}=\frac{(-1)^q}{2^{2M-1}}\cdot$$
$$\cdot\frac{i}{\prod_{\substack{1\le k\le 2M\\ k\ne p}}\sin\frac{\phi_p-\phi_k}{2}}\left\{\left(V_{M+q}\overline{V}_M-V_{M+q-1}\overline{V}_{M-1}\right)+\left(V_{M-q}\overline{V}_M-V_{M-q-1}\overline{V}_{M-1}\right)\right\}\cdot$$
$$\cdot\frac{1}{\sum_M\cos\left\{\frac{1}{2}(\phi_1+\ldots+\phi_{2M})-(\phi_1+\ldots+\phi_{2M})_M\right\}} \tag{3.82}$$

If we introduce the following symbols:

$C_t = e^{\frac{1}{2}(\phi_1+\ldots+\phi_{p-1}+\phi_{p+1}+\ldots+\phi_{2M})i}\overline{V_t} = A_t + iB_t$

$A_t^{(p)} = A_t = \sum_t \cos\left\{\frac{1}{2}(\phi_1+\ldots+\phi_{p-1}+\phi_{p+1}+\ldots+\phi_{2M}) - (\phi_1+\ldots+\phi_{p-1}+\phi_{p+1}+\ldots+\phi_{2M})_t\right\}$

$B_t^{(p)} = B_t = \sum_t \sin\left\{\frac{1}{2}(\phi_1+\ldots+\phi_{p-1}+\phi_{p+1}+\ldots+\phi_{2M}) - (\phi_1+\ldots+\phi_{p-1}+\phi_{p+1}+\ldots+\phi_{2M})_t\right\} \Rightarrow$

$(V_{M+q}\overline{V}_M - V_{M+q-1}\overline{V}_{M-1}) + (V_{M-q}\overline{V}_M - V_{M-q-1}\overline{V}_{M-1}) = \overline{C}_{M+q}C_M - \overline{C}_{M+q-1}C_{M-1}+$
$+\overline{C}_{M-q}C_M - \overline{C}_{M-q-1}C_{M-1}$

$\overline{C}_{M+q} = A_{M+q} - iB_{M+q}$

$A_{M+q} = A_{M-q-1};\ B_{M+q} = -B_{M-q-1} \Rightarrow$

$\overline{C}_{M+q} = C_{M-q-1};\ \overline{C}_{M+q-1} = C_{M-q};\ C_{M-1} = A_{M-1} + iB_{M-1};\ A_{M-1} = A_M;\ B_{M-1} = -B_M \Rightarrow$

$(V_{M+q}\overline{V}_M - V_{M+q-1}\overline{V}_{M-1}) + (V_{M-q}\overline{V}_M - V_{M-q-1}\overline{V}_{M-1}) = A_M(C_{M-q-1} - C_{M-q} + \overline{C}_{M-q} - \overline{C}_{M-q-1})$
$+ iB_M(C_{M-q-1} + C_{M-q} + \overline{C}_{M-q} + \overline{C}_{M-q-1}) =$
$= 2iA_M(B_{M-q-1} - B_{M-q}) + 2iB_M(A_{M-q-1} + A_{M-q}) \quad (3.83)$

It follows that:

$$\frac{\mathbf{X}_p^q}{\mathbf{X}_{2M}} = \frac{(-1)^{q+1}}{2^{2M-2}} \cdot \frac{A_M^{(p)}\left(B_{M-q-1}^{(p)} - B_{M-q}^{(p)}\right) + B_M^{(p)}\left(A_{M-q-1}^{(p)} + A_{M-q}^{(p)}\right)}{\prod\limits_{\substack{1\le k\le 2M \\ k\ne p}} \sin\frac{\phi_p-\phi_k}{2} \sum_M \cos\left\{\frac{1}{2}(\phi_1+\ldots+\phi_{2M}) - (\phi_1+\ldots+\phi_{2M})_M\right\}};$$
$$for\ 1 \le q \le M-2 \quad (3.84)$$

In addition, for:

$$\begin{aligned} &q = M-1 \Rightarrow V_0 = 1;\ C_0 = e^{\frac{1}{2}(\varphi_1+\ldots+\varphi_{p-1}+\varphi_{p+1}+\ldots+\varphi_{2M})i} \\ &A_0 = \cos\frac{1}{2}(\varphi_1+\ldots+\varphi_{p-1}+\varphi_{p+1}+\ldots+\varphi_{2M}); \\ &B_0 = \sin\frac{1}{2}(\varphi_1+\ldots+\varphi_{p-1}+\varphi_{p+1}+\ldots+\varphi_{2M}) \\ &q = M \Rightarrow V_{-1} = 0 \Rightarrow C_{-1} = 0; \\ &A_{-1} = B_{-1} = 0\ (A_0\ and\ B_0\ are\ the\ same\ as\ before) \end{aligned} \quad (3.85)$$

Now, we can determine co-factors $\mathbf{F}_p^q$ for $1 \le p \le 2M;\ M+1 \le q \le 2M$, and $\mathbf{F}_p^{M+q}$ for $1 \le q \le M$. As above, we obtain that:

$$\mathbf{F}_p^q = \frac{(-1)^{\frac{M(M+1)}{2}}}{2^M} e^{-(M-1)\frac{\pi}{2}i} e^{-M(\varphi_1+\ldots+\varphi_{p-1}+\varphi_{p+1}+\ldots+\varphi_{2M})i}$$
$$\left(\Delta_{2M+1,M+1}^{(p,M+q)} - \Delta_{2M+1,M+1}^{(p,M-q+1)}\right)_{x_j=e^{\varphi_j i}} \quad (3.86)$$

From this, it follows that:

$$\frac{\mathbf{X}_p^{M+q}}{\mathbf{X}_{2M}} = \frac{(-1)^{q+1}}{2^{2M-2}} \cdot \\ \cdot \frac{A_M^{(p)}\left(A_{M-q}^{(p)} - A_{M-q-1}^{(p)}\right) + B_M^{(p)}\left(B_{M-q}^{(p)} + B_{M-q-1}^{(p)}\right)}{\prod\limits_{\substack{1\le k\le 2M \\ k\ne p}} \sin\frac{\varphi_p-\varphi_k}{2} \cdot \sum_M \cos\left\{\frac{1}{2}(\varphi_1 + \ldots + \varphi_{2M}) - (\varphi_1 + \ldots + \varphi_{2M})_M\right\}} \tag{3.87}$$

We obtain:

$$\mathbf{F}_p^q = \frac{(-1)^{\frac{M(M+1)}{2}}}{2^M} e^{-M\frac{\pi}{2}i} e^{-M(\varphi_1+\ldots+\varphi_{p-1}+\varphi_{p+1}+\ldots+\varphi_{2M})i} \\ \left(\Delta_{2M+1,M+1}^{(p,M+q)} - \Delta_{2M+1,M+1}^{(p,M-q+1)}\right)_{x_j=e^{\varphi_j i}} \\ for\ 1 \le p \le 2M \wedge 1 \le q \le M \tag{3.88}$$

$$\mathbf{F}_p^q = \frac{(-1)^{\frac{M(M+1)}{2}}}{2^M} e^{-(M-1)\frac{\pi}{2}i} e^{-M(\varphi_1+\ldots+\varphi_{p-1}+\varphi_{p+1}+\ldots+\varphi_{2M})i} \\ \left(\Delta_{2M+1,M+1}^{(p,M+q)} - \Delta_{2M+1,M+1}^{(p,M-q+1)}\right)_{x_j=e^{\varphi_j i}} \tag{3.89}$$

for $1 \le p \le 2M;\ M+1 \le q \le 2M$, and $\mathbf{F}_p^{M+q}$ for $1 \le q \le M$.

Using the derived analytical relation, the unknown parameters of the signal (amplitude, phase) can be determined by dividing the expression representing the solution of the appropriate co-determinants by the expression representing the analytical solution to the system determinant.

$$\psi_k = arctg\frac{\mathbf{X}_{M+k}}{\mathbf{X}_k} \\ a_k = \frac{1}{A_k\mathbf{X}}\sqrt{\mathbf{X}_k^2 + \mathbf{X}_{M+k}^2} \tag{3.90}$$

To separate the DC component from the input signal during the processing, a low-frequency filter can be employed. By measuring this separated component before subjecting the signal to differentiation, the proposed reconstruction system in this study enables the processing of a wide range of periodic input signals.

Similar to previous cases, it is necessary to know or assume the order of the highest M harmonic component in the processed signal spectrum, even if it exceeds the expected value. To estimate the frequencies of multi (single)-sinusoidal signals, we utilized modified techniques described in Eqs. (1.7) and (1.8), which are based on the detection of zero-crossings (ZCs) in the processed signal. Figure 3.12 illustrates the root mean squared errors (RMSE) of the frequency estimates (f) and the

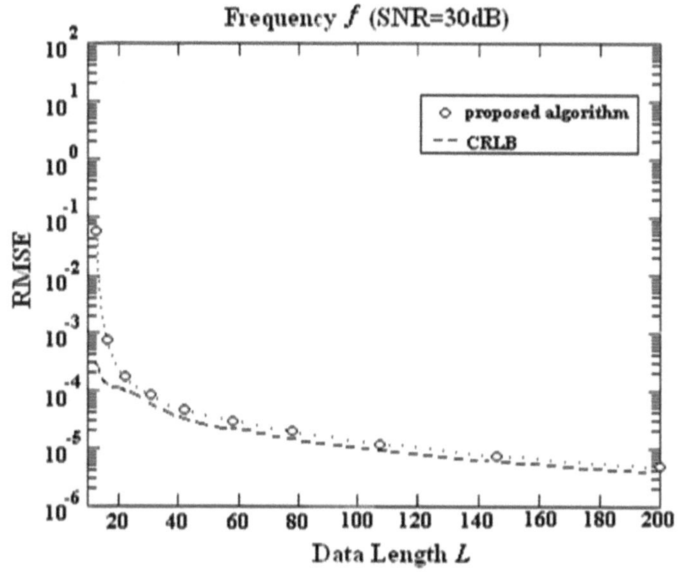

Fig. 3.12 RMSE of the estimates of f and the associated CRLB versus L when SNR = 30 dB

corresponding Cramer-Rao lower bound (CRLB) as a function of the data length (L), given a signal-to-noise ratio (SNR) of 30 dB.

However, in scenarios where the signal-to-noise ratio is very low and accompanied by significant frequency modulation (FM), it is possible to modify the estimation procedure without imposing additional requirements on the implementation, as presented in [22, 23, 42]. Additionally, alternative solutions can be derived from the approaches described in [17], where accurate frequency estimation algorithms for multiple real sinusoids in white noise, based on the linear prediction approach, were developed. The first algorithm minimizes the weighted least squares (WLS) cost function, subject to a generalized unit-norm constraint, while the second method is a WLS estimator with a monic constraint. Both algorithms provide close frequency estimates with accuracies approaching the Cramer-Rao lower bound for white Gaussian noise. Another modified parameter estimator based on a magnitude phase-locked loop principle was proposed in [18], which demonstrated improved tracking capabilities when the fundamental component of the signal became small or temporarily disappeared.

To recalculate the unknown parameters (amplitude and phase) of the processed periodic signals, the results of the differentiation of the input analog signals $x(t_l)$ (Eq. (3.58)) are required. The differentiation circuit consists of an oversampling analog-to-digital conversion unit with a dithering process and a higher-order finite-impulse response (FIR) digital differentiator, followed by a decimator as proposed in [43]. The differential samples of the input signal are obtained by differentiating the signal at precisely defined time moments, which are determined with respect to the detected zero-crossing moments.

The values of the derived relations depend on the measured frequency f, as the determinant elements are calculated based on the coefficient α_l, as shown in Eq. (3.59). Therefore, it is necessary to recalculate the frequency of the carrier signal, considering possible changes, when recalculating the new variables A_k and φ_l. This adjustment reduces the potential errors in the reconstruction process caused by variations in the frequency of the processed signal.

The parameters of the derived system of equations are independent of the starting moment of differentiation (sampling) of the input signal. The moments t_l at which differentiation occurs are completely random (asynchronous) and unrelated to the frequency of the processed signal. These moments primarily depend on the speed of the differentiation circuit, the sample-and-hold (S/H) circuit, and the analog-to-digital conversion circuit, which together form a numerical equivalent of the input signal's derivative.

Figure 3.13 demonstrates the impact of the error in determining the frequency of the carrier signal on the relative error in determining the value of the system determinant for various harmonic contents of the input periodic signal. Meanwhile, the A_k coefficients exhibit less dependence on the fundamental signal frequencies, as shown in Fig. 3.14. For systems requiring high-speed calibration procedures, an uncertainty analysis similar to the one conducted in [25] can be performed. In such cases, the effect of uncertainty introduced by the time base generator (jitter) can be modeled as non-stationary additive noise. The paper [25] also presents a method to calculate an uncertainty bound around the reconstructed waveform based on a desired confidence level. The error resulting from the assumed non-idealities in the proposed reconstruction model falls within the boundaries specified by [25, 26]. A sensitivity function, commonly formulated assuming noise-free data, provides point-wise information about the reliability of the reconstructed signal prior to sampling, as described in [26], where the minimum error bound of signal reconstruction is derived considering noisy data.

The time required to perform the necessary differentiations of the input signal for reconstruction purposes is denoted by $2Mt_{sample}$. This time represents an approximation of the overall reconstruction time in simulations. In practical applications, the reconstruction time needs to be increased to account for the estimation of variables $x(t_l)$, A_l and α_l, which is directly dependent on the value of N, as well as the time interval Δt required for other recalculations according to the proposed algorithm. Therefore, the total reconstruction time can be defined as $N \cdot 2M \cdot t_{sample} + \Delta t \approx N/f + \Delta t$, taking into consideration the necessary synchronization with the zero crossing of the input signal.

To evaluate the performance of the proposed algorithm, computer simulations are conducted using input data. Statistical properties of the estimator are investigated by utilizing noisy samples generated through computer simulation. Noisy samples are obtained by adding white noise samples to the processed signal. The Cramer-Rao lower bound (CRLB), which is the inverse of the Fisher information, is commonly used as a lower bound for the mean squared error (MSE) in estimating deterministic

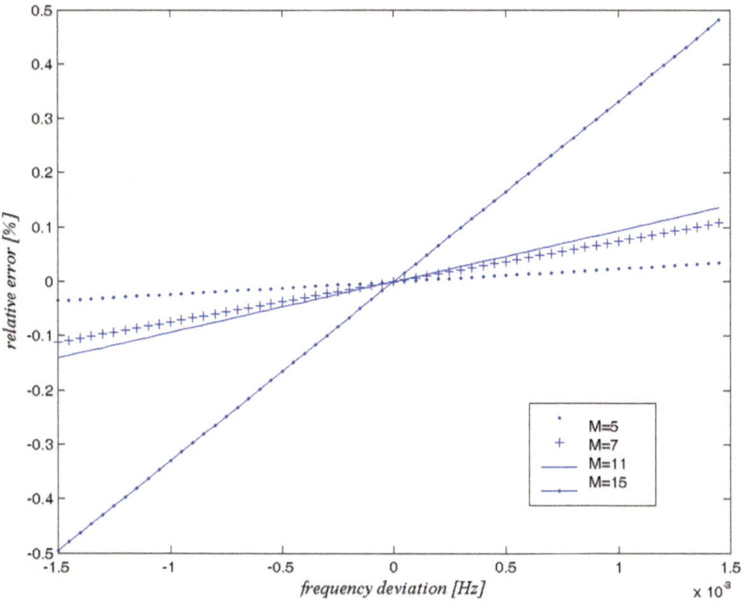

Fig. 3.13 Relative error in calculation of the system determinant as function of error in synchronisation with frequency of fundamental harmonic of the input signal

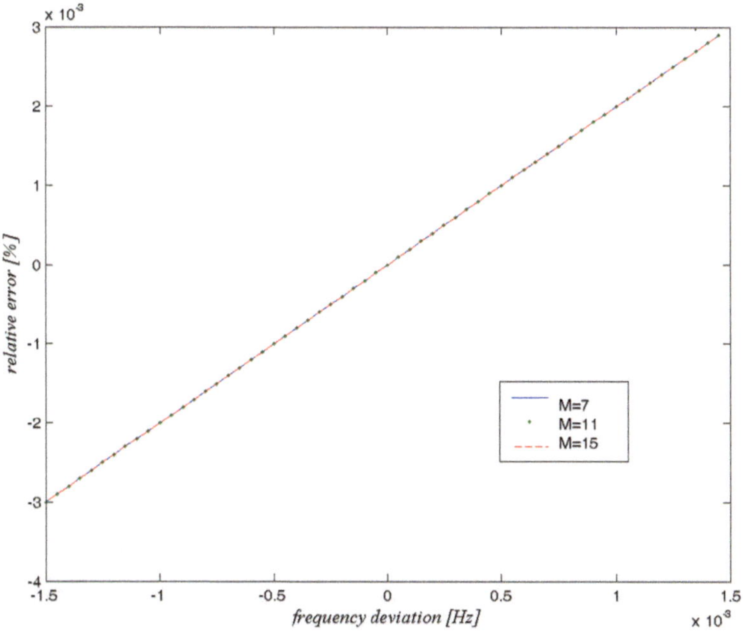

Fig. 3.14 Relative error in determination of variables A_k as function of error in synchronisation with frequency of fundamental harmonic of the input signal

parameters [44–46]. Figures 3.15 and 3.16 illustrate the MSE of the amplitudes and frequency, respectively, after 105 simulations. The results clearly demonstrate that the proposed estimation scheme asymptotically approaches the CRLB, similar to the findings in [46, 47].

The impact of noise on the signal is further studied by estimating the magnitudes of signals contaminated with noise. A sinusoidal 50-Hz test signal with additive white centered Gaussian noise is used as the input for the test. The random noise is selected to achieve a specified signal-to-noise ratio (SNR), defined as SNR = $20\log\left(A/\sqrt{2}\sigma\right)$, where A represents the magnitude of the signal's fundamental harmonics, and s is the standard deviation of the noise. Figure 3.17 presents the

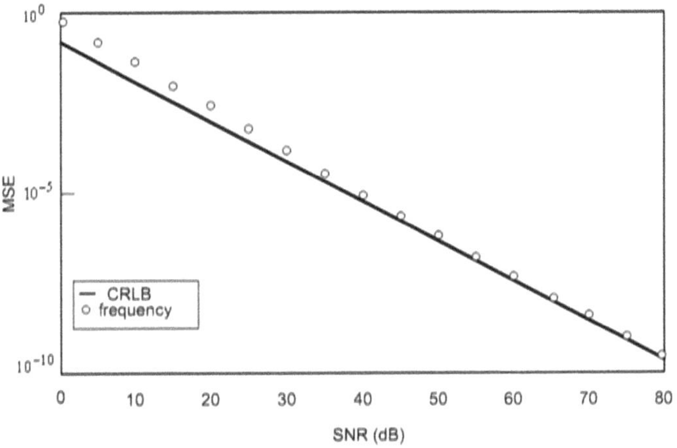

Fig. 3.15 MSE of the frequency as a function of SNR

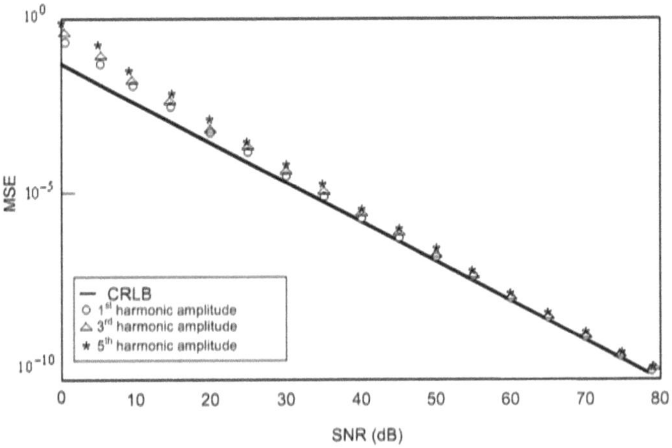

Fig. 3.16 MSE of the three harmonic amplitudes as a function of SNR

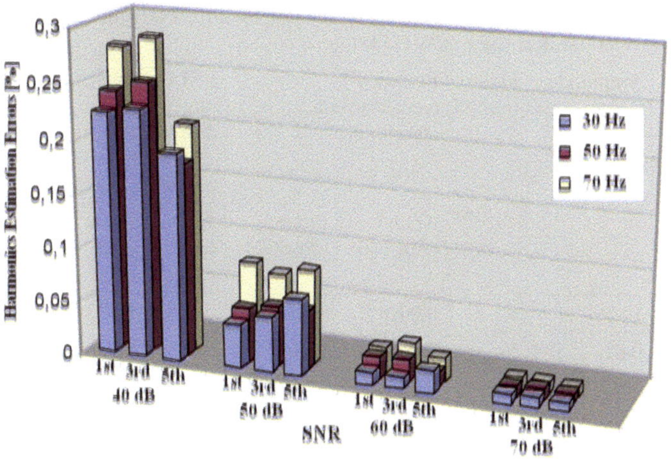

Fig. 3.17 Maximum estimation errors for noisy input signals

maximum errors observed in frequency and harmonic magnitude estimates when input signals of 30, 50, and 70 Hz with SNRs of 40, 50, 60, and 70 dB are used. It should be noted that in practical scenarios, the SNR of voltage signals obtained from power systems typically ranges between 50 and 70 dB. With the proposed technique, only a few errors are expected at this noise level, as depicted in Fig. 3.17.

The proposed algorithm was further tested using simulation in the Matlab and SIMULINK software packages. A standard sigma-delta ADC with an effective resolution of 24 bits and a sampling rate of 1 kHz was employed as the ADC model. The parameters of the input signal used in the simulation corresponded to the values provided in Table 3.6. The execution time of the proposed algorithm on the specified hardware platform was measured to be 0.0167 s. During the simulation, the output Power Spectral Density (PSD) of the ideal signal, as well as the signals affected by thermal noise and clock jitter, was observed. The PSD values ranged from −100 to −170 dB, corresponding to Signal-to-Noise Distortion Ratios (SNDR) between 55 and 76 dB. These results demonstrate the algorithm's effectiveness in maintaining high signal quality and low distortion levels, even in the presence of noise and jitter.

A signal comprising the first 7 harmonics, with a fundamental frequency of 50 Hz, was utilized for the analysis. In the simulated environment, the presence of noise and jitter introduced a relative error of 0.01 in the detection of the fundamental frequency.

To facilitate a comprehensive comparison between the various estimation procedures, Fig. 3.18 displays the Power Spectral Density (PSD) of a 1-s segment of the processed signal, with the parameters specified in Table 3.6. To mitigate the leakage effect during the Fast Fourier Transform (FFT), the Hanning window was applied.

The performance of the proposed algorithm demonstrates accuracy comparable to existing methods for processing signals of similar nature, as reported in [30, 34, 37, 49, 50], and surpasses the results presented in [48]. In the time domain, the

Table 3.6 Comparison of simulation results by the proposed reconstruction algorithm, FFT and continuous wavelet transformation (CWT) [48]

Harmonic number	Amplitude (Vpp)	Phase (rad)	Proposed reconstruction algorithm		FFT (sampling rate = 25 kHz; data length = 25,000; time period = 1 s)		CWT	
			Amp. error (%)	Phase error (%)	Amp. error (%)	Phase error (%)	Amp. error (%)	Phase error (%)
1	1	π	0.0018	0.0019	0.296	0.322	0.023	0.034
2	0.81	π/3	0.0024	0.0022	0.035	0.038	0.032	0.028
3	0.62	0	0.0022	0.0021	0.875	0.843	0.049	0.026
4	0.58	π/6	0.0015	0.0016	0	0	0.144	0.012
5	0.41	π/4	0.0015	0.0021	0	0	0.013	0.154
6	0.33	π/12	0.0021	0.0019	0	0	0.012	0.017
7	0.16	0	0.0020	0.0022	0	0	0.223	0.186

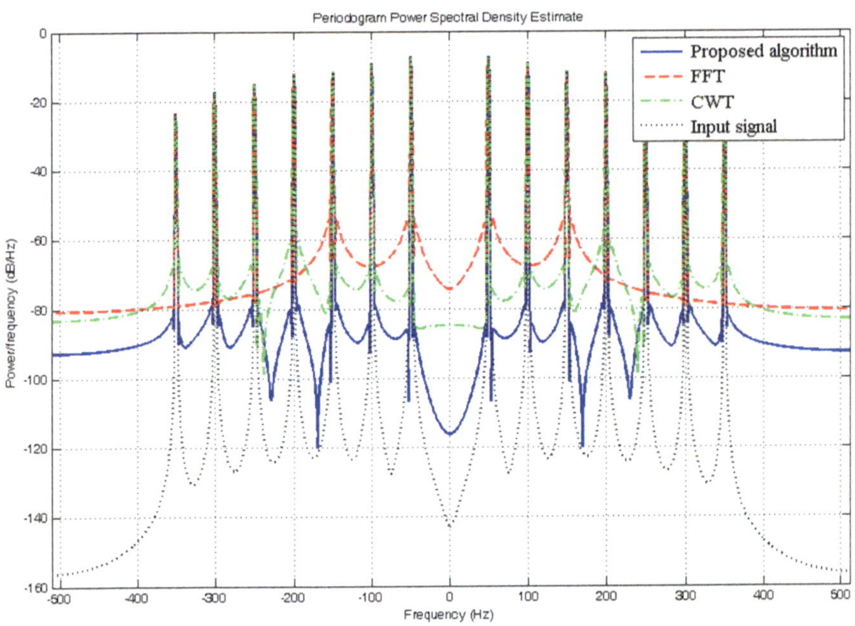

Fig. 3.18 Comparison among the different estimation respect to techniques applied, with PSD

relative error between the reconstructed signal and the original signal was found to be 0.0025%. The observed errors in amplitude and phase detection primarily stem from inaccuracies in measuring the input signal samples and determining the values of the derived equations.

3.6 New Estimation Procedure Based on Usage of Finite-Impulse-Response Comb Filters and Digital Differentiator

In contrast to existing algorithms for parameter estimation in power quality monitoring and harmonic compensation, algorithm [51] introduces a novel approach that enables simultaneous estimation of the fundamental frequency, as well as the amplitudes and phases of harmonic waves. By utilizing finite-impulse-response (FIR) comb filters, the algorithm transforms a multi-harmonic input signal into a pure sinusoidal signal. Its core methodology relies on the utilization of partial derivatives of the processed signal and a weighted estimation procedure to accurately estimate the fundamental frequency, amplitude, and phase of the multi-sinusoidal signal.

The algorithm [51] incorporates several key components within its framework. Firstly, an oversampling analog-to-digital conversion unit is employed, which includes a dithering process. This unit is followed by a finite-impulse-response (FIR) comb filter and a higher-order finite-impulse-response (FIR) digital differentiator, which in turn is connected to a decimator. This arrangement ensures effective signal processing and parameter estimation.

The proposed method is particularly suited for scenarios where the input signal's fundamental frequency is limited to the bandwidth of the first harmonic component. It distinguishes itself from existing parameter estimation algorithms by enabling simultaneous estimation of signal parameters while accommodating time-varying frequencies.

Simulation results validate the effectiveness of the proposed algorithm, highlighting its potential for precise measurements of significant electrical quantities such as RMS measurements of periodic signals, power, and energy.

In order to obtain a single sinusoidal signal from the input signal described in Eq. (3.1), it is necessary for the frequency response of the filters to have nulls at the expected harmonic frequencies and unity gain at the fundamental frequency. If the frequency is not constant, the filter parameters must be adapted online during the frequency estimation process. In order to achieve accurate measurements, it becomes essential to track the system frequency and apply specific corrections to the measurement algorithms and input filters.

Figure 3.19 depicts the block diagram of the adaptive algorithm that utilizes finite-impulse-response (FIR) comb filters. These FIR comb filters, as described in Ref. [52], consist of second-order modules that eliminate the DC component and harmonic frequencies, while maintaining unity gain at the fundamental frequency.

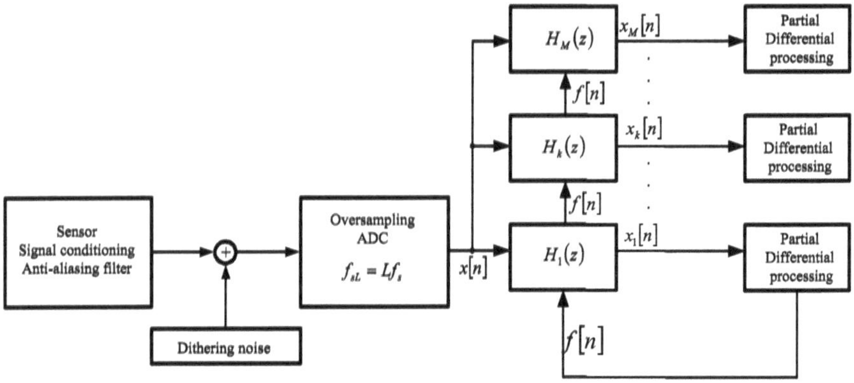

Fig. 3.19 Block diagram of the estimation algorithm with FIR comb filters

The complete filter is constructed by cascading these individual modules. The second-order section, which rejects the DC component and the frequency $f_S/2$ (where f_S represents the sampling frequency) and has unity gain at the frequency of the kth harmonic $f_k = kf_1$, can be represented by the following transfer function:

$$H_{0k}(z) = \frac{1 - z^{-2}}{\left|1 - z_k^{-2}\right|} \tag{3.91}$$

where $\left|1 - z_k^{-2}\right| = 2\sin(k\omega_1 T)$, $z^{-1} = e^{-j\omega}$, $\omega = 2\pi f/\omega_S$, $z_1^{-1} = e^{-j\omega_1}$, $\omega_1 = 2\pi f_1/\omega_S$, f_1 is the fundamental frequency (ω_S is the sampling frequency), and $z_k^{-1} = e^{-jk\omega_1}$.

The transfer function representing the section that eliminates the harmonics $\omega_i = i\omega_1$, while maintaining unity gain at the frequency $\omega_k = k\omega_1$, can be expressed as follows:

$$H_{ik}(z) = \frac{1 - 2\cos(\omega_i T)z^{-1} + z^{-2}}{\left|1 - 2\cos(\omega_i T)z_k^{-1} + z_k^{-2}\right|} \tag{3.92}$$

where $\left|1 - 2\cos(\omega_i T)z_k^{-1} + z_k^{-2}\right| = 2|\cos(k\omega_1 T) - \cos(i\omega_1 T)|$, $i = 1, 2, 3, \ldots, M$, $i \neq k$ represents the gain adjustment for the kth harmonic. $M = [f_S/2f_1]$ is calculated as the largest integer less than or equal to $f_S/2f_1$, which corresponds to the number of sections in the cascade. The transfer function of the filter for the kth harmonic is given as:

$$H_k(z) = H_{0k}(z)\prod_{\substack{i=1\\ i\neq k}}^{M} H_{ik}(z) \tag{3.93}$$

The transfer function (3.93) for the filter exhibits an uncontrolled phase shift at the fundamental frequency. In certain applications, such as reactive power measurement, an additional phase shift of $\pi/2$ needs to be introduced. This can be achieved through the use of an adaptive phase shifter [52]. The proposed estimation algorithm is characterized by its simplicity, as it employs closed-form solutions for calculating the filter coefficients. Moreover, the number of sections in the cascade and the lengths of the data window can be adjusted during the measurement process to accommodate frequency changes.

It should be noted that the filters exhibit non-unity gains at frequencies other than the nominal power system frequency. Therefore, adaptation of the filter response during the estimation process is essential to achieve unity gain. Additionally, the incorporation of the finite-impulse-response (FIR) filter in this modified discrete Fourier transform (DFT) approach allows for the suppression of all signal harmonics that can cause the leakage effect. The FIR filter, being adaptive in nature, adjusts its coefficients based on the estimation of the actual frequency. The accuracy of the overall algorithm depends on the precision of the frequency estimation.

The estimation of multi-sine wave signal frequencies from a finite set of noisy discrete-time measurements is a significant task both theoretically and practically. Extensive research has been conducted on this problem, which remains an active area of study to this day [53–57]. It finds applications in various fields, including control theory, relaying protection, intelligent instrumentation of power systems, signal processing, digital communications, distribution automation, biomedical engineering, radar applications, radio frequency, and instrumentation and measurement. Numerous algorithms have been developed, such as the adaptive notch filter, time–frequency representation-based methods, phase-locked loop-based methods, eigen-subspace tracking estimation, extended Kalman filter frequency estimation, and internal model-based methods. The choice of the frequency estimator depends on the specific requirements of the application, which may include factors such as accuracy, processing speed or complexity, and the ability to handle multiple signals.

By differentiating the kth harmonic component of the signal (3.1) after filtering (as depicted in Fig. 3.19), the following expression is obtained:

$$\frac{d(x_k(t))}{dt} = \frac{d(X_k \sin(k\omega t + \varphi_k))}{dt}\Big|_{t=t_n} = y_{1k}(t_n) = y_{1k}[n] \Rightarrow$$
$$y_{1k}(t_n) = k\omega X_k \cos(k\omega t_n + \varphi_k) \quad (3.94)$$

$$\frac{d^2(x_k(t))}{dt^2} = \frac{d^2(X_k \sin(k\omega t + \varphi_k))}{dt^2}\Big|_{t=t_n} = y_{2k}(t_n) = y_{2k}[n] \Rightarrow$$
$$y_{2k}(t_n) = -k^2\omega^2 X_k \sin(k\omega t_n + \varphi_k) \quad (3.95)$$

In the context of this analysis, t_n represents the arbitrary time instant at which the differentiation of the input analog signal takes place, allowing for irregularly spaced samples. Figure 2.1 illustrates a schematic diagram illustrating the process of obtaining the first and second-order derivatives within an oversampling system.

Using the differentially obtained values in the form of samples, the unknown parameters of the signal can be determined through the following calculations:

$$
\begin{aligned}
f_k(t_n) &= kf(t_n) = \frac{1}{2\pi}\sqrt{\left|\frac{y_{2k}(t_n)}{x_k(t_n)}\right|} \\
\varphi_k(t_n) &= arctg\left(\frac{x_k(t_n)}{y_{1k}(t_n)}2k\pi f(t_n)\right) - 2k\pi f(t_n)t_n \\
X_k(t_n) &= \frac{x_k(t_n)}{\sin(2k\pi f(t_n)t_n + \varphi_k(t_n))}
\end{aligned}
\tag{3.96}
$$

As depicted in Fig. 3.19, the signal of interest is acquired by a sensor, which undergoes conditioning through a signal conditioning circuit (amplifier) and is band-limited by an anti-aliasing filter. Subsequently, the conditioned analog signal $x(t)$ is combined with dithering noise, enabling the resulting composite signal to be fed into an analog-to-digital converter (ADC) unit at an oversampling rate of $f_{sL} = Lf_s$ Hz (samples/s). Here, f_s represents the minimum sampling rate (Nyquist sampling rate), and L denotes the oversampling factor. Each digital sample $x_k[n]$, which is encoded with N_q bits, is obtained.

The first and second-order derivatives of the digitized signal are then computed using first and second-order finite-impulse-response (FIR) digital differentiators operating at the oversampling rate. These differentiators possess transfer functions designed as $H_{1D}(z)$ and $H_{2D}(z)$, as illustrated in Fig. 2.1. The anti-aliasing filter depicted in Fig. 3.19 has a bandwidth of $f_s/2$ Hz.

Despite the fact that adding dithering noise raises the average spectral noise floor of the original input signal, the dithering process disrupts the coherence between the quantization error and the original input signal. Consequently, the spectrum of the quantization noise becomes white and flat. Thus, the oversampling technique can be effectively employed to compensate for the reduced signal-to-quantization-noise ratio (SQNR) and further enhance the SQNR by increasing the sampling rate. Typically, the random wideband dithering noise, which can be generated by a noise diode or noise generator integrated circuits (ICs), exhibits a root-mean-square (RMS) level equivalent to 1/3 to 1 least significant bit (LSB) voltage level.

The performance evaluation of algorithm [51] was conducted through computer simulations using input data. The input data consisted of a multi-sine signal subjected to first and second derivative calculations, employing an 8-bit analog-to-digital converter (ADC) resolution and an oversampling factor L of 256. Prior to over-sampling, dithering noise $\sigma_d = (1/3)LSB$ was added to the analog signal. This resulted in oversampling the multi-sine signal with a total of $4096{\cdot}L$ samples.

In Fig. 3.20, the maximum errors in frequency and harmonic magnitude estimates are displayed for different scenarios. Specifically, input signals with frequencies of 30, 50, and 70 Hz were utilized, with signal-to-noise ratios (SNRs) ranging from 40 to 70 dB. The figure illustrates that, even with noise at these levels, the proposed technique exhibits minimal error in the estimated parameters, as indicated by the results presented in Fig. 3.20.

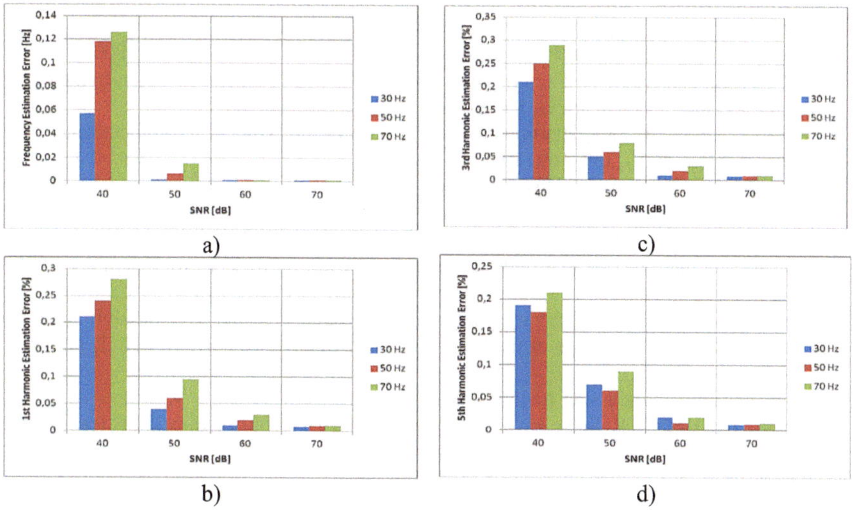

Fig. 3.20 Maximum estimations errors for noisy input signals: **a** frequency; **b** 1st harmonic; **c** 3rd harmonic; **d** 5th harmonic

References

1. Proakis, J.G., Manolakis, D.G.: Digital Signal Processing: Principles, Algorithms, Applications, 3rd edn. Prentice-Hall, Englewood Cliffs, NJ (1996)
2. Prendergast, R.S., Levy, B.C., Hurst, P.J.: Reconstruction of band-limited periodic nonuniformly sampled signals through multirate filter banks. IEEE Trans. Circuits Syst.-I **51**(8), 1612–1622 (2004)
3. Marziliano, P., Vetterli, M., Blu, T.: Sampling and exact reconstruction of bandlimited signals with additive shot noise. IEEE Trans. Inf. Theory **52**(5), 2230–2233 (2006)
4. Margolis, E., Eldar, Y.C.: Reconstruction of nonuniformly sampled periodic signals: algorithms and stability analysis. In: Proceedings of the 2004 11th IEEE International Conference on Electronics, Circuits and Systems, 2004 (ICECS 2004), pp. 555–558, 13–15 Dec 2004
5. Sun, W., Zhou, X.: Reconstruction of band-limited signals from local averages. IEEE Trans. Inf. Theory **48**(11), 2955–2963 (2002)
6. Muciek, A.K.: A method for precise RMS measurements of periodic signals by reconstruction technique with correction. IEEE Trans. Instrum. Meas. **56**(2), 513–516 (2007)
7. Petrovic, P.: New approach to reconstruction of nonuniformly sampled AC signals. In: Proceedings of 2007 IEEE International Symposium on Industrial Electronics (ISIE 2007), Vigo, Spain, pp. 1693–1698 (2007)
8. Bos, A.V.D.: Estimation of Fourier coefficients. IEEE Trans. Instrum. Meas. **38**(5), 1005–1007 (1989)
9. Bos, A.V.D.: Estimation of complex Fourier coefficients. IEE Proc.-Control Theory Appl. **142**(3), 253–256 (1995)
10. Pintelon, R., Schoukens, J.: An improved sine-wave fitting procedure for characterizing data acquisition channels. IEEE Trans. Instrum. Meas. **45**(2), 588–593 (1996)
11. Xiao, Y., Tadokaro, Y., Shida, K.: Adaptive algorithm based on least mean p-power error criteterion for Fourier analysis in additive noise. IEEE Trans. Signal Proc. **47**(4), 1172–1181 (1999)
12. Petrovic, P., Stevanovic, M.: Algorithm for Fourier coefficient estimation. IET Signal Proc. **5**(2), 138–149 (2001)

13. Agrawal, M., Prasad, S., Roy, C.D.: A simple solution for the analytic inversion of Van der Monde and Confluent Van der Monde matrices. IETE J. Res. **47**(5), 217–219 (2001)
14. Gohberg, I., Olshevsky, V.: The fast generalized Parker-Traub algorithm for inversion of Van der Monde and related matrices. J. Complex. **13**(2), 208–234 (1997)
15. Jog, C.S.: On the explicit determination of the polar decomposition in n-dimensional vector spaces. J. Elast. **66**(2), 159–169 (2002)
16. Neagoe, V.E.: Inversion of the Van der Monde matrix. IEEE Signal Process. Lett. **3**(4), 119–120 (1996)
17. So, H.C., Chan, K.W., Chan, Y.T., Ho, K.C.: Linear prediction approach for efficient frequency estimation of multiple real sinusoids: algorithms and analyses. IEEE Trans. Signal Proc. **53**(7), 2290–2305 (2005)
18. Wu, B., Bodson, M.: Frequency estimation using multiple source and multiple harmonic components. In: Proceedings of the 2002 American Control Conference, vol. 1, pp. 21–22 (2002)
19. Seber, G.: Linear Regression Analysis. Wiley, New York (1977)
20. Reeves, S.J., Heck, L.P.: Selection of observations in signal reconstruction. IEEE Trans. Signal Proc. **43**(3), 788–791 (1995)
21. Poberezhskiy, Y.S., Poberezhskiy, G.Y.: Sampling and signal reconstruction circuits performing internal antialiasing filtering and their influence on the design of digital receivers and transmitters. IEEE Trans. Circuits Syst.-I **51**(1), 118–129 (2004)
22. Xiao, Y., Ward, R.K., Ma, L., Ikuta, A.: A new LMS-based Fourier analyzer in the presence of frequency mismatch and applications. IEEE Trans. Circuits Syst.-I **52**(1), 230–245 (2005)
23. Xiao, Y., Ward, R.K., Xu, L.: A new LMS-based Fourier analyzer in the presence of frequency mismatch. In: Proc. 2003 Int. Symp. Circuits and Systems 2003 (ISCAS'03), vol. 4, pp. 369–372
24. Petrovic, P.: New digital multimeter for accurate measurement of synchronously sampled AC signals. IEEE Trans. Instrum. Meas. **53**(3), 716–725 (2004)
25. Daboczi, T.: Uncertainty of signal reconstruction in the case of jitter and noisy measurements. IEEE Trans. Instrum. Meas. **47**(5), 1062–1066 (1998)
26. Wang, G., Han, W.: Minimum error bound of signal reconstruction. IEEE Signal Proc. Lett. **6**(12), 309–311 (1999)
27. Nigham, N.J.: Accuracy and Stability of Numerical Algorithms, 2nd edn. SIAM (2002)
28. Feichtinger, H.G.: Reconstruction of band-limited signals from irregular samples, a short summary. In: 2nd International Workshop on Digital Image Processing and Computer Graphics with Applications, pp. 52–60 (1991)
29. Cooklev, T.: An efficient architecture for orthogonal wavelet transforms. IEEE Signal Proc. Lett. **13**(2), 77–79 (2006)
30. Schoukens, J., Rolain, Y., Simon, G., Pintelon, R.: Fully automated spectral analysis of periodic signals. IEEE Trans. Instrum. Meas. **52**(4), 1021–1024 (2003)
31. Reeves, S.J.: An efficient implementation of the backward greedy algorithm for sparse signal reconstruction. IEEE Signal Proc. Lett. **6**(10), 266–268 (1999)
32. Coakley, K.J., Wang, C.M., Hale, PD., Clement, T.S.: Adaptive characterization of jitter noise in sampled high-speed signals. IEEE Trans. Instrum. Meas. **52**(5), 1537–1547 (2003)
33. Hoseini, H.Z., Kale, I., Shoaei, O.: Modeling of switched-capacitor delta-sigma modulators in SIMULINK. IEEE Trans. Instrum. Meas. **54**(4), 1646–1654 (2005)
34. Hidalgo, R.M., Fernandez, J.G., Rivera, R.R., Larrondo, H.A.: A simple adjustable window algorithm to improve FFT measurements. IEEE Trans. Instrum. Meas. **51**(1), 31–36 (1996)
35. Petrovic, P.: Frequency and parameter estimation of multi-sinusoidal signal. Meas. Sci. Rev. **12**(5), 175–183 (2012)
36. Petrovic, P.: New method and circuit for processing of band-limited periodic signals. SIViP **6**(1), 109–123 (2012)
37. Petrovic, P., Marjanovic, S., Stevanovic, M.: Measuring of slowly changing AC signals without sample and hold circuit. IEEE Trans. Instrum. Meas. **49**(6), 1245–1248 (2000)
38. Vendersteen, G., Pintelon, R.: Maximum likelihood estimator for jitter noise models. IEEE Trans. Instrum. Meas. **49**(6), 1282–1284 (2000)

39. Stenbakken, G.N., Deyst, J.P.: Timebase distortion measurements using multiphase sinewaves. In: IEEE Instrum. Meas. Techn. Conf., Ottawa, Canada, pp. 1003–1008, May 1997
40. Petrovic, P.: New procedure for estimation of amplitude and phase of analog multiharmonic signal based on the differential irregular samples. J. Signal Process. Syst. **81**(1), 11–27 (2015)
41. Petrovic, P.: Fourier coefficient estimation based on the differential values of the processed signal. IETE J. Res. **56**(1), 30–43 (2011)
42. Vizireanu, D.N.: A simple and precise real-time four point single sinusoid signals instantaneous frequency estimation method for portable DSP based instrumentation. Measurement **44**(2), 500–502 (2011)
43. Tan, L., Wang, L.: Oversampling technique for obtaining higher order derivative of low-frequency signals. IEEE Trans. Instrum. Meas. **60**(11), 3677–3684 (2011)
44. Kay, S.M.: Modern Spectral Estimation: Theory and Applications. Prentice-Hall, Englewood Cliffs, NJ (1988)
45. Stoica, P., Li, H., Lim, J.: Amplitude estimation of sinusoidal signals: survey, new results, and an application. IEEE Trans. Signal Process. **48**(2), 338–352 (2000)
46. Belega, D., Dallet, D., Slepicka, D.: Accurate amplitude estimation of harmonic components of incoherently sampled signals in the frequency domain. IEEE Trans. Instrum. Meas. **59**(5), 1158–1166 (2010)
47. Pantazis, Y., Roces, O., Stylianou, Y.: Iterative estimation of sinusoidal signal parameters. IEEE Signal Process. Lett. **17**(5), 461–464 (2010)
48. Tse, N.C.F., Lai, L.L.: Wavelet-based algorithm for signal analysis. EURASIP J. Adv. Signal Process. **2007**, 38916 (2007)
49. Agrež, D.: Weighted multi-point interpolated DFT to improve amplitude estimation of multi-frequency signal. IEEE Trans. Instrum. Meas. **51**, 287–292 (2002)
50. Agrež, D.: Improving phase estimation with leakage minimisation. IEEE Trans. Instrum. Meas. **54**(4), 1347–1353 (2005)
51. Petrovic, P.: Algorithm for simultaneous parameters estimation of multi-harmonic signal. Metrol. Meas. Syst. **19**(4), 693–702 (2012)
52. Kuseljevic, M.D.: A simple method for design of adaptive filters for sinusoidal signals. IEEE Trans. Instrum. Meas. **57**(10), 2242–2249 (2008)
53. Wu, J., Long, J., Wang, J.: High-accuracy, wide-range frequency estimation methods for power system signals under nonsinusoidal conditions. IEEE Trans. Power Deliv. **20**(1), 366–374 (2005)
54. So, H.C., Chan, K.W., Chan, Y.T., Ho, K.C.: Linear prediction approach for efficient frequency estimation of multiple real sinusoids: algorithms and analyses. IEEE Trans. Signal Process. **53**(7), 2290–2305 (2005)
55. Klein, J.D.: Fast algorithms for single frequency estimation. IEEE Trans. Signal Process. **54**(5), 1762–1770 (2006)
56. El-Shafey, M.H., Mansour, M.M.: Application of a new frequency estimation technique to power systems. IEEE Trans. Power Deliv. **21**(3), 1045–1053 (2006)
57. Trapero, J.R., Sira-Ramirez, H., Batlle, V.F.: An algebraic frequency estimator for a biased and noisy sinusoidal signal. Signal Process. **87**(6), 1188–1201 (2007)

Chapter 4
New Procedure for Harmonics Estimation Based on Hilbert Transformation

In practical scenarios, spectrum estimation of discretely sampled processes often relies on the fast Fourier transform (FFT). While FFT is highly effective when the sampling frequency and fundamental frequency are synchronized, it fails to provide satisfactory performance when desynchronization and non-stationary conditions are present. The accuracy of FFT deteriorates under these conditions, particularly when the fundamental/harmonic frequency varies over time. Such errors arise due to the different magnitude gains of the orthogonal finite-impulse-response (FIR) filters at frequencies other than the nominal power frequency. This issue is exacerbated by the fact that the frequencies of harmonics align with the zero points of the frequency response of the FIR filter with a rectangular window, commonly employed in DFT algorithms. In practice, short data records are often analyzed, amplifying these problems inherent to measurement procedures. Sliding discrete Fourier transform (SDFT) algorithms offer efficient computation of DFT in a sliding mode, requiring only a few multiplications and additions per computation, with complexity independent of the signal length.

In real-world scenarios, it is challenging to ensure coherent frequency relationships between all frequencies in the input signal and the sampling frequency, leading to the well-known leakage phenomena. To mitigate this undesired effect, the "windowing" procedure is commonly employed. Additionally, the Interpolated DFT (IpDFT) method is used to estimate signal parameters under non-coherent sampling conditions in the frequency domain. Although highly precise, this method entails a substantial computational burden.

Efforts have been made to enhance the periodicity requirement of the FFT process by employing time-weighting functions or correction interpolation algorithms. While these approaches reduce potential calculation errors, complete elimination is not achieved. Synchronization with the grid fundamental frequency is necessary unless a windowing technique is utilized. Traditional synchronization methods like phase-locked loops (PLL) exhibit long response times, particularly when dealing with transient phenomena in the input signal, such as power supply frequency variations or

P. Petrović, *Processing, Estimation and Measurement of Signals Parameters in Public Distribution Networks*, Studies in Systems, Decision and Control 502,
https://doi.org/10.1007/978-3-031-43107-4_4

phase jumps. Furthermore, FFT suffers from disadvantages related to frame implementation, as it processes entire frames of data and cannot provide in-between data. Sliding mode FFT calculations demand intensive computational effort.

To address the sensitivity of algorithms to frequency variations, a recursive Newton-type algorithm is proposed in [1]. This algorithm simultaneously estimates the spectrum and frequency of the power signal, overcoming the challenges associated with frequency variation. The proposed approach adopts a non-linear signal model, necessitating non-linear estimation, and employs a recursive formulation. By employing a sequential tuning strategy with a forgetting factor, the algorithm achieves superior convergence and accuracy.

FIR filters with optimized frequency response do not necessitate synchronization between generators and acquisition devices, except when the least-squares (LS) technique [2] is employed during their design. However, this approach demands greater computational power compared to situations where synchronization is achieved. It is well-established that the LS design method for large-order filters entails significant computational requirements [3], surpassing what can be accomplished within a single sampling interval. Consequently, these filters are not suitable for online applications, especially considering the need for adaptation to potential deviations in the processing signal frequency.

The Hilbert-Huang transform (HHT) represents a novel signal analysis method [4, 5]. HHT is derived from the principles of empirical mode decomposition (EMD) and Hilbert transform (HT). Through EMD, complex data sets can be decomposed into a finite number of intrinsic mode functions (IMFs), enabled by the well-behaved Hilbert transform. Unlike wavelet methods, which rely on wavelet base selection for wavelet transform results, the HHT method is inherently self-adaptive. It eliminates the need for time base selection and offers more stable outcomes.

The work in [6] explores the application of Hilbert Transform (HT) signal processing techniques and matrix calculations for estimating harmonics in power systems. Assuming that the signal's fundamental frequency has already been estimated independently, new analytical relations are derived to enable fast calculation of the unknown parameters with low numerical error. This method exhibits potential for successful implementation in various signal processing applications such as signal reconstruction, spectral estimation, and measurement and monitoring in power systems.

In this approach [6], the instantaneous attributes of processed signals are obtained using the analytic signal (AS) approach of Gabor, employing discrete HT. The mathematical foundation of these algorithms is first reviewed, followed by a description of the method's practical implementation using a time-domain approach. The resulting system of equations can be solved simply using the derived analytical and summarized expressions. A comparison of numerical results obtained with the proposed method against established techniques is provided, demonstrating the strong appeal of the proposed procedure for online estimation of observed power signal parameters. Furthermore, the proposed algorithm allows for estimation of parameters with minimal numerical error and relatively low processor power requirements.

4.1 Analytical Signals and Fourier Coefficients

In 1946, Gabor [7] introduced the concept of complex signals represented by analytical functions, wherein the real and imaginary parts constitute a Hilbert transform pair. Consider a real-valued signal denoted as $x(t)$, and let $H(x(t))$ represent the Hilbert transform of $x(t)$, which is defined as follows:

$$\mathcal{H}(x(t)) = -\frac{1}{\pi}\, p.v. \int_{-\infty}^{+\infty} \frac{x(\tau)}{\tau - t}\, d\tau, \tag{4.1}$$

where *p.v.* denotes Cauchy principal value of the integral. Then complex-valued signal whose imaginary part is Hilbert transform of its real part:

$$u_a(t) = x(t) + i \cdot \mathcal{H}(x(t)) \tag{4.2}$$

is called *analytical signal.* Notice that following equations hold:

$$\begin{aligned} x(t) &= \mathrm{Re}\{u_a(t)\} = \frac{u_a(t) + \overline{u_a(t)}}{2} \quad \mathit{and} \\ H(x(t)) &= \mathrm{Im}\{u_a(t)\} = \frac{u_a(t) - \overline{u_a(t)}}{2} \end{aligned} \tag{4.3}$$

where $\overline{u_a(t)}$ denotes the complex conjugate of $u_a(t)$.

If the signal $x(t)$ can be expressed as a sinusoidal Fourier series limited to the first M harmonic components (with the fundamental frequency f), then:

$$x(t) = a_0 + \sum_{k=1}^{M} a_k \sin(2k\pi f t + \psi_k) \tag{4.4}$$

then its Hilbert transform pair is:

$$\mathcal{H}(x(t)) = \sum_{k=1}^{M} a_k \cos(2k\pi f t + \psi_k) \tag{4.5}$$

and thus, the corresponding analytical signal, i.e., the complex fixed version of input signal $x(t)$, is given by:

$$u_a(t) = a_0 + \sum_{k=1}^{M} a_k \cdot e^{i(2k\pi f t + \psi_k)} \tag{4.6}$$

By sampling the signal $u_a(t)$, we obtain $M + 1$ equations of the form:

$$u_a(t_l) = a_0 + \sum_{k=1}^{M} a_k \cdot e^{i(2k\pi f t_l + \psi_k)}, l = 1, 2, \ldots, M + 1, \tag{4.7}$$

where t_l is the sampling time moment.

To estimate the unknowns, which include the amplitudes and phases of the M harmonics, as well as the DC value of the input signal, a system of equations is constructed following the same form as Eq. (4.7). This system can be expressed as:

$$u_a(t_l) = a_0 + \sum_{k=1}^{M} a_k e^{j2k\pi f t_l} \cdot \Psi_k, l = 1, 2, \ldots, M + 1 \tag{4.8}$$

where $\Psi_k = e^{j\psi_k}$ for $k = 1, 2, \ldots, M$.

Consider the determinant of the following system of equations:

$$\Delta = \begin{vmatrix} a_0 & a_1 e^{j2\pi f t_1} & \cdots & a_M e^{j2M\pi f t_1} \\ a_0 & a_1 e^{j2\pi f t_2} & \cdots & a_M e^{j2M\pi f t_2} \\ \vdots & \vdots & & \vdots \\ a_0 & a_1 e^{j2\pi f t_{M+1}} & \cdots & a_M e^{j2M\pi f t_{M+1}} \end{vmatrix}$$
$$= a_0 a_1 \cdots a_M \begin{vmatrix} 1 & e^{j2\pi f t_1} & \cdots & e^{j2M\pi f t_1} \\ 1 & e^{j2\pi f t_2} & \cdots & e^{j2M\pi f t_2} \\ \vdots & \vdots & & \vdots \\ 1 & e^{j2\pi f t_{M+1}} & \cdots & e^{j2M\pi f t_{M+1}} \end{vmatrix} \tag{4.9}$$

This determinant can be represented as:

$$\Delta = a_0 a_1 \cdots a_M \cdot V_{M+1}\left(e^{j2\pi f t_1}, e^{j2\pi f t_2}, \ldots, e^{j2\pi f t_{M+1}}\right) \tag{4.10}$$

where V_{M+1} is Van der Monde determinant of type $M + 1$. Thus, we have:

$$\Delta = a_0 a_1 \cdots a_M \prod_{1 \le p < q \le M+1} \left(e^{j2\pi f t_q} - e^{j2\pi f t_p}\right) \tag{4.11}$$

Extensive analysis has been dedicated to the determinant mentioned above [8, 9]. However, the existing literature primarily focuses on proposing program-based procedures, often iterative in nature, to determine the value of the original and inverted Van der Monde's matrix with improved efficiency. In contrast, the proposed concept introduces Hilbert's transformation into the procedure, thereby distinguishing it from the approach presented in [10].

Let Δ_k represent a determinant obtained from Δ by replacing the kth column with another column.

$$\left[u_a(t_1)\ u_a(t_2) \cdots u_a(t_{M+1})\right]^T, \text{i.e.,} \tag{4.12}$$

$$\Delta_k = a_0 \cdots a_{k-1} a_{k+1} \cdots a_M \cdot \begin{vmatrix} 1 & \ldots & e^{j2(k-1)\pi f t_1} & u_a(t_1) & e^{j2(k+1)\pi f t_1} & \ldots & e^{j2M\pi f t_1} \\ 1 & \ldots & e^{j2(k-1)\pi f t_2} & u_a(t_2) & e^{j2(k+1)\pi f t_2} & \ldots & e^{j2M\pi f t_2} \\ \vdots & & \vdots & \vdots & \vdots & & \vdots \\ 1 & \cdots & e^{j2(k-1)\pi f t_{M+1}} & u_a(t_{M+1}) & e^{j2(k+1)\pi f t_{M+1}} & \cdots & e^{j2M\pi f t_{M+1}} \end{vmatrix} \tag{4.13}$$

Further on, let Δ_{kl} be the algebraic cofactor corresponding to element $u_a(t_l)$ of determinant Δ_k. Then we have:

$$\Delta_k = \sum_{l=1}^{M+1} u_a(t_l) \cdot \Delta_{kl} \tag{4.14}$$

and consequently the algebraic cofactor corresponding to element $u_a(t_l)$ of determinant Δ_k is:

$$\Delta_{kl} = \begin{bmatrix} a_0 \cdots & a_{k-1} e^{j2(k-1)\pi f t_1} & a_{k+1} e^{j2(k+1)\pi f t_1} & \cdots & a_M e^{j2M\pi f t_1} \\ \vdots & \vdots & \vdots & & \vdots \\ a_0 \cdots & a_{k-1} e^{j2(k-1)\pi f t_{l-1}} & a_{k+1} e^{j2(k+1)\pi f t_{l-1}} & \cdots & a_M e^{j2M\pi f t_{l-1}} \\ a_0 \cdots & a_{k-1} e^{j2(k-1)\pi f t_{l+1}} & a_{k+1} e^{j2(k+1)\pi f t_{l+1}} & \cdots & a_M e^{j2M\pi f t_{l+1}} \\ \vdots & \vdots & \vdots & & \vdots \\ a_0 \cdots & a_{k-1} e^{j2(k-1)\pi f t_{M+1}} & a_{k+1} e^{j2(k+1)\pi f t_{M+1}} & \cdots & a_M e^{j2M\pi f t_{M+1}} \end{bmatrix} \tag{4.15}$$

$$\Delta_{kl} = a_0 \cdots a_{k-1} a_{k+1} \cdots a_M \cdot \begin{vmatrix} 1 \cdots & e^{j2(k-1)\pi f t_1} & e^{j2(k+1)\pi f t_1} & \cdots & e^{j2M\pi f t_1} \\ \vdots & \vdots & \vdots & & \vdots \\ 1 \cdots & e^{j2(k-1)\pi f t_{l-1}} & e^{j2(k+1)\pi f t_{l-1}} & \cdots & e^{j2M\pi f t_{l-1}} \\ 1 \cdots & e^{j2(k-1)\pi f t_{l+1}} & e^{j2(k+1)\pi f t_{l+1}} & \cdots & e^{j2M\pi f t_{l+1}} \\ \vdots & \vdots & \vdots & & \vdots \\ 1 \cdots & e^{j2(k-1)\pi f t_{M+1}} & e^{j2(k+1)\pi f t_{M+1}} & \cdots & e^{j2M\pi f t_{M+1}} \end{vmatrix} \tag{4.16}$$

Let:

$$\begin{aligned} & e^{j2\pi f t_l} + \cdots + e^{j2k\pi f t_l} + \cdots + e^{j2M\pi f t_l} = -p_1 \\ & e^{j2\pi f t_l} e^{j4\pi f t_l} + e^{j2\pi f t_l} e^{j6\pi f t_l} \cdots + e^{j2(M-1)\pi f t_l} e^{j2M\pi f t_l} = p_2 \\ & \cdots \\ & e^{j2\pi f t_l} \cdots e^{j2(M-1)\pi f t_l} + \cdots + e^{j4\pi f t_l} \cdots e^{j2M\pi f t_l} = (-1)^{M-1} p_{M-1} \\ & e^{j2\pi f t_l} \cdots e^{j2M\pi f t_l} = (-1)^M p_M \end{aligned} \tag{4.17}$$

If we introduce following notation (for every complex number z):

$$P(z;l) = \prod_{m=1,\, m\neq l}^{M+1} \left(z - e^{j2k\pi f t_m}\right) = z^M + p_1 z^{M-1} + \cdots + p_{M-1} z + p_M \tag{4.18}$$

It's follows that $P(e^{j2k\pi f t_l}; l) = 0$. Then, for every $m = 1, 2, \ldots, l-1, l+1, \ldots, M+1$ holds:

$$e^{j2M\pi f t_m} = -p_1 e^{j2(M-1)\pi f t_m} - \cdots - p_{M-1} e^{j2\pi f t_m} - p_M \tag{4.19}$$

and therefore:

$$\Delta_{kl} = a_0 \cdots a_{k-1} a_{k+1} \cdots a_M \cdot \begin{vmatrix} 1 \cdots & e^{j2(k-1)\pi f t_1} & e^{j2(k+1)\pi f t_1} & \cdots & -p_1 e^{j2(M-1)\pi f t_1} - \cdots - p_{M-1} e^{j2\pi f t_1} - p_M \\ \vdots & \vdots & \vdots & & \vdots \\ 1 \cdots & e^{j2(k-1)\pi f t_{l-1}} & e^{j2(k+1)\pi f t_{l-1}} & \cdots & -p_1 e^{j2(M-1)\pi f t_{l-1}} - \cdots - p_{M-1} e^{j2\pi f t_{l-1}} - p_M \\ 1 \cdots & e^{j2(k-1)\pi f t_{l+1}} & e^{j2(k+1)\pi f t_{l+1}} & \cdots & -p_1 e^{j2(M-1)\pi f t_{l+1}} - \cdots - p_{M-1} e^{j2\pi f t_{l+1}} - p_M \\ \vdots & \vdots & \vdots & & \vdots \\ 1 \cdots & e^{j2(k-1)\pi f t_{M+1}} & e^{j2(k+1)\pi f t_{M+1}} & \cdots & -p_1 e^{j2(M-1)\pi f t_{M+1}} - \cdots - p_{M-1} e^{j2\pi f t_{M+1}} - p_M \end{vmatrix} \tag{4.20}$$

$$\Delta_{kl} = a_0 \cdots a_{k-1} a_{k+1} \cdots a_M \cdot \begin{vmatrix} 1 \cdots & e^{j2(k-1)\pi f t_1} & e^{j2(k+1)\pi f t_1} & \cdots & -p_{M-k} e^{j2k\pi f t_1} \\ \vdots & \vdots & \vdots & & \vdots \\ 1 \cdots & e^{j2(k-1)\pi f t_{l-1}} & e^{j2(k+1)\pi f t_{l-1}} & \cdots & -p_{M-k} e^{j2k\pi f t_{l-1}} \\ 1 \cdots & e^{j2(k-1)\pi f t_{l+1}} & e^{j2(k+1)\pi f t_{l+1}} & \cdots & -p_{M-k} e^{j2k\pi f t_{l+1}} \\ \vdots & \vdots & \vdots & & \vdots \\ 1 \cdots & e^{j2(k-1)\pi f t_{M+1}} & e^{j2(k+1)\pi f t_{M+1}} & \cdots & -p_{M-k} e^{j2k\pi f t_{M+1}} \end{vmatrix} \tag{4.21}$$

$$\Delta_{kl} = a_0 \cdots a_{k-1} a_{k+1} \cdots a_M \cdot (-p_{M-k}) \cdot (-1)^{M-k-1}. \begin{vmatrix} 1 \cdots & e^{j2(k-1)\pi f t_1} & e^{j2k\pi f t_1} & e^{j2(k+1)\pi f t_1} & \cdots & e^{j2M\pi f t_1} \\ \vdots & \vdots & \vdots & \vdots & & \vdots \\ 1 \cdots & e^{j2(k-1)\pi f t_{l-1}} & e^{j2k\pi f t_{l-1}} & e^{j2(k+1)\pi f t_{l-1}} & \cdots & e^{j2M\pi f t_{l-1}} \\ 1 \cdots & e^{j2(k-1)\pi f t_{l+1}} & e^{j2k\pi f t_{l+1}} & e^{j2(k+1)\pi f t_{l+1}} & \cdots & e^{j2M\pi f t_{l+1}} \\ \vdots & \vdots & \vdots & \vdots & & \vdots \\ 1 \cdots & e^{j2(k-1)\pi f t_{M+1}} & e^{j2k\pi f t_{M+1}} & e^{j2(k+1)\pi f t_{M+1}} & \cdots & e^{j2M\pi f t_{M+1}} \end{vmatrix} \tag{4.22}$$

$$\Delta_{kl} = (-1)^{M-k} p_{M-k} a_0 \cdots a_{k-1} a_{k+1} \cdots a_M \cdot V_M\left(e^{j2\pi f t_1}, \ldots, e^{j2k\pi f t_{l-1}}, e^{j2k\pi f t_{l+1}}, \ldots, e^{j2k\pi f t_{M+1}}\right) \tag{4.23}$$

$$\Delta_k = \sum_{l=1}^{M+1} (-1)^{M-k} u_a(t_l) p_{M-k} a_0 \cdots a_{k-1} a_{k+1} \cdots a_M \prod_{\substack{1 \le p < q \le M+1 \\ p,q \ne l}} \left(e^{j2\pi f t_q} - e^{j2\pi f t_p}\right) \tag{4.24}$$

Now, we have:

$$\Psi_k = e^{j\psi_k} = \frac{\Delta_k}{\Delta} = \frac{1}{a_k} \cdot z_k \tag{4.25}$$

Furthermore, we have:

$$a_k \cos \psi_k = \mathrm{Re} z_k \quad and \quad a_k \sin \psi_k = \mathrm{Im} z_k \tag{4.26}$$

which implies:

$$a_k = \sqrt{\{\mathrm{Re} z_k\}^2 + \{\mathrm{Im} z_k\}^2} \quad and \quad \psi_k = arctg \frac{\mathrm{Im} z_k}{\mathrm{Re} z_k} \tag{4.27}$$

4.1.1 Description of Algorithm

Input $x(t)$

Step 1. For $l = 1, 2, \ldots, M+1$, sampling $x(t_l)$;

Step 2. Hilbert transform of $x(t_l)$;

Step 3. Analytical signal $u_a(t_l) = x(t_l) + iH(x(t_l))$;

Step 4. Compute Δ,

    for $k = 0, 1, 2, \ldots, M$,

        for $l = 1, 2, \ldots, M+1$,

            compute Δ_{kl},

    compute Δ_k and compute Ψ_k and z_k;

  compute a_k and ψ_k.

Step 5 Output a_k and ψ_k

Once the amplitudes and phase values of each harmonic component in the input signal have been determined as described earlier, it becomes possible to define the instantaneous frequency as follows:

$$f_k = \frac{1}{2\pi} \frac{d(\psi_k)}{dt} \tag{4.28}$$

Thus, providing an additional verification of the frequency spectrum of the processed signal, these values can be utilized as input data for a subsequent least-squares (LS) fitting procedure. This additional step can further refine the estimation of various signal parameters, particularly the frequency estimate.

In contrast to [11], which also employs Hilbert's transformation for subsequent estimation of harmonic components in periodic signals, the analysis conducted in this example focuses on transforming the obtained Van der Monde-type matrix. This transformation facilitates the derivation of novel analytical and concise expressions that allow for exact solutions of the system of equations (4.8) when this matrix is utilized as the starting point of the analysis. Consequently, the authors have successfully avoided the need for an additional estimation procedure, as observed in [10, 11]. This advancement has led to the development of an algorithm that enables highly precise recalibration of parameters in the processed multi-harmonic power signals.

4.2 Proposed Estimation Algorithm

For the proposed algorithm, it is necessary to know or determine in advance the order of the highest M harmonic component in the spectrum of the processed signal. It is acceptable to choose an M value that is greater than the expected (actual) value.

According to the proposed processing concept, the frequency of the fundamental harmonic in the processed signal can be either known beforehand or determined before forming the analytical signal based on measurements of the real signal.

Extensive literature exists on the topic of frequency measurement in various power system applications. Papers [12–16] provide descriptions of recursive methods for local system frequency measurement based on three consecutive samples. In [13, 15], the estimation problem is improved by transforming it into an overdetermined set of linear equations and incorporating the least-squares (LS) algorithm. Weighted LS (WLS) algorithm with a suitable forgetting factor tuning procedure is used in [12, 14]. Median filtering is employed to identify and ignore bad data corresponding to poor frequency estimates. One advantage of median filtering is that there is no need to specify a cutoff frequency [17]. One of the mentioned methods can be utilized to estimate the frequency spectrum of the processed signals.

Based on the procedure described in the previous chapter, it is evident that the moment at which the sampling of the input signal begins is completely independent of the frequency of the processed signal, as well as the sampling frequency (as defined in Eq. (4.7)). The moments t_l primarily depend on the speed of the analog-to-digital

(AD) conversion circuit and can be defined as $t_l = lt_{sample}$, where $t_{sample} = 1/f_S$, (f_S represents the sampling frequency). Implementing sampling and reconstruction with internal anti-aliasing filtering significantly improves the performance of digital receivers [3, 18], allowing for more accurate reconstruction with reduced errors.

Figure 4.1 illustrates the process of generating an analytic signal using a Hilbert transformer.

The Hilbert transformer, known as a quadrature filter or wide-band 90° phase filter, is a linear two-port network that produces the Hilbert transform of the input signal at its output. In the discrete domain, the impulse response $h(n)$ of a Hilbert transformer can be defined as [19, 20]:

$$h(n) = \begin{cases} \frac{2}{n\pi}\sin^2\left(\frac{n\pi}{2}\right) & for \quad n \neq 0 \\ 0 & for \quad n = 0 \end{cases} \tag{4.29}$$

The impulse response of a Hilbert transformer is not absolutely summable. However, it is possible to approximate the ideal Hilbert transformer using IIR or FIR systems, making Hilbert transformers a special class of filters. IIR Hilbert transformers exhibit phase and magnitude errors when approximating the ideal frequency response. These filters can be designed by utilizing two all-pass systems with phase responses differing by approximately $\pi/2$ within a defined frequency band of $0 < |\omega| < \pi$. By taking the real and imaginary parts of the complex signal as the outputs of these two all-pass filters, it can be observed that the spectrum of such a signal nearly vanishes in a significant portion of the negative frequency range. Hence, the outputs of the two all-pass filters closely resemble a Hilbert transformer [19].

FIR Hilbert transformers with constant group delay are relatively straightforward to design. They accurately achieve a $\pi/2$ phase shift, with an additional linear phase component required for causality in the FIR system. Evaluating the impulse response (4.29) for both positive and negative values of n reveals that it exhibits anti-symmetry. Consequently, FIR Hilbert transformers are based on either Type III (odd-length L with anti-symmetric impulse response) or Type IV (even-length L with anti-symmetric impulse response) symmetry. Type III filters have zero amplitude at $\omega = 0$ and $\omega = \pi$, while Type IV filters only have zero amplitude at $\omega = 0$. Therefore, the FIR approximation is suitable within a specified range of frequencies (the pass-band region), excluding these extreme values [19].

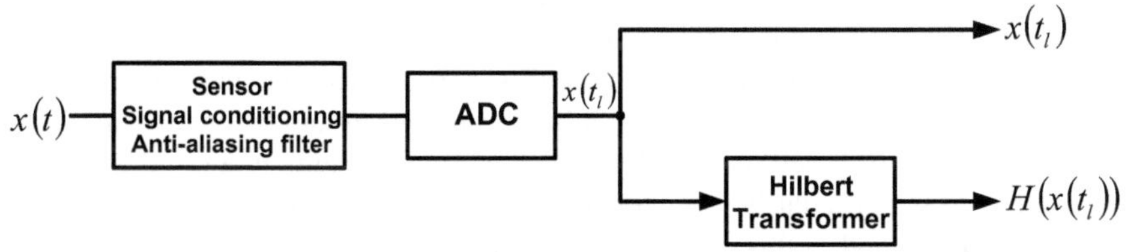

Fig. 4.1 Generation of an analytic signal using a Hilbert transformer

The precise phase characteristics of Type III and Type IV FIR systems make them desirable for approximating Hilbert transformers. Furthermore, while IIR Hilbert transformers can suffer from instability and sensitivity to coefficient rounding errors, FIR filters offer guaranteed stability and are less affected by coefficient rounding. The phase response of FIR filters remains unaffected by such rounding errors. Given that the phase requirement is fulfilled in FIR Hilbert transformers, designing a FIR Hilbert transformer involves finding the impulse response $h(n)$ for $n = 0$ to $L - 1$, satisfying the specified magnitude response requirement [19].

$$(1-\delta) \le \left|H\left(e^{j\omega}\right)\right| \le (1-\delta) \quad for \quad \omega_L \le \omega \le \omega_H \tag{4.30}$$

The magnitude response specification of a Hilbert transformer is defined by the allowed pass-band ripple δ and the frequency response $H(e^{j\omega})$. The lower pass-band edge ω_H is determined as $\omega_H = \pi - \omega_L$, for a Type III Hilbert transformer or $\omega_H = \pi$ for a Type IV Hilbert transformer. By increasing the length L of the filter, the values of ω_L and ω_H can be made to approach 0 and π, respectively. The transition band is considered as $\omega_L/2\pi$ for Hilbert transformers. To design optimum equiripple FIR linear phase Hilbert transformers, the Parks-McClellan algorithm, as utilized in our proposed algorithm, is commonly employed [19, 21]. The order of the filter, denoted as $L - 1$, needs to be estimated in advance. An estimation formula for L is provided in [21, 22], and in line with this, we utilize an FIR filter of order 100. For a Type III Hilbert transformer, the number of multipliers, denoted as m, is given by $(L + 1)/4$ if $L = 4k + 3$ or $(L - 1)/4$ if $L = 4k + 1$, where k is an integer. In the latter case, the impulse response values $h(0)$ and $h(L - 1)$ are both zero. For a Type IV Hilbert transformer, the number of multipliers is $L/2$ [19].

4.2.1 Numerical Complexity of Proposed Algorithm

The algorithm presented in this paper, used to calculate the unknown parameters of the processed signal, exhibits a computational complexity of approximately $O((M)^2)$ flops in its implementation. This is in contrast to standard matrix procedures integrated into software packages such as Matlab and Mathematica, which require $O((2M)^3)$ flops [23], where M represents the number of harmonic coefficients.

It is well-known that the complexity of iterative procedures based on Fast Fourier Transforms (FFTs) remains unchanged as the number of sampling values increases in certain non-matrix implementations. This allows for improved convergence of such algorithms. However, increasing the number of samples in standard matrix methods, whether iterative or based on pseudo-inverse matrices, results in a rapid growth in complexity [23], particularly when processing signals with large spectra. In contrast, the proposed method in this paper operates under an opposite scenario. It does not require the determination of the inversion matrix, leading to a faster solution with better convergence compared to other matrix-based methods. The reconstruction algorithm employed in this paper utilizes certain properties of the Van der Monde matrix, without the need for its explicit inversion. Calculating high powers of the coefficients, typically problematic in single or even double precision calculations, is avoided. Additionally, the proposed method is non-iterative, making it faster to implement in practical applications compared to other matrix procedures. Unlike FFT, where precision improves with an increase in the number of samples, the number of samples required by the proposed algorithm is equal to the number of unknown values in the system of equations defined in (4.8), with the sampling frequency being independent of the frequency of the processed signal.

The algorithm proposed in this paper exhibits greater stability compared to the standard method [24] and is free from propagation errors. In the standard procedure, errors that occur during the calculation of the base harmonic's amplitude propagate throughout the method as they are used to determine the parameters of other harmonic components in order to reconstruct the processed signal. Consequently, errors in the determination of frequency and amplitude at early stages propagate to higher harmonics, leading to contamination of the determination of the nth harmonic by the errors in phase and amplitude from previous steps [3].

4.2.2 Computing Time

The proposed algorithm is suitable for operation with sigma-delta ADC, allowing for high resolution and speed in processing input signals. The time required to perform the necessary number of signal samplings for reconstruction can be approximated as $(M + 1)t_{sample}$, which represents the time needed for harmonic estimation in simulations. In practical applications, the estimation time should be increased to account for the Hilbert transformation of the sampled input signal, estimation of variables a_k and ψ_k (using an additional LS or WLS fitting procedure), and the time interval Δt needed for other recalculations according to the proposed algorithm. Consequently, the total reconstruction time can be defined as $P(M + 1)t_{sample} + \Delta t$ (P passages is arbitrary). The speed of the proposed algorithm is comparable to the algorithms analyzed in [25, 26].

To demonstrate the efficiency of the new procedure, a comparison is made of the relative execution times of the proposed algorithm, FFT, and the continuous wavelet transform (CWT) [27]. The comparisons were performed using the Matlab program package (version R2016a) on a computer with an Intel Core i5-4460 CPU, 16 GB RAM, and Windows 7 operating system. FFT was executed at a sampling rate of 25 kHz, with a data length of $N = 25{,}000$ and a time period of 1 s. The same sampling rate was used for CWT, employing the complex Morlet wavelet as the mother wavelet. The number of data points per cycle of any harmonic was maintained at a minimum of 25 for accurate amplitude detection [27]. Higher sampling frequencies would yield better harmonic amplitude detection but require more data points, resulting in slower computation. To expedite CWT computation, the simulated signal was down-sampled for the detection of lower harmonics. As per the classical uncertainty principle, larger time windows are required at low frequencies, while smaller time windows suffice at high frequencies. The numerical complexity of FFT is O(Nlog$N$), whereas CWT with scaling factors s (the dilation parameter or scale of the wavelet) poses a complexity of O(Nlog$N$xs) [27, 28]. The execution time for the CWT algorithm is the longest because the number of scales is too high to achieve good frequency resolution. With a more powerful hardware platform and a different programming environment, the execution time for the proposed algorithm will be significantly shorter (Table 4.1).

The paper [29] provides a measurement of the required processor time for implementing the matrix method in signal reconstruction, as commonly implemented in many program packages. In contrast, the method proposed in this paper does not necessitate any special memorization of the transformation matrix or recalculation of the inversion matrix. Consequently, it is more efficient in implementation and not limited to sparse matrices. Moreover, the proposed solution lends itself to easier hardware realization, and the algorithm can be practically implemented on any platform.

The proposed procedure reduces the time needed to solve the system of equations (4.8) by 3–4 times compared to the Gaussian elimination with partial pivoting (GEPP) method offered in the Matlab program package, depending on the number of harmonic components in the processed signal. All calculations are performed using IEEE standard double floating-point arithmetic with a unit round-off of $u \approx 1.1 \times 10^{-16}$. The difference in the obtained values and results obtained with the GEPP algorithm was on the order of 1×10^{-14}.

This algorithm can also be applied for spectral analysis, enabling determination of the amplitude and phase values of signal harmonics based on the prescribed system of equations. By following a step-by-step approach in conducting the described procedure, it is possible to accurately identify the spectral content and subsequently optimize the proposed algorithm accordingly. The accuracy of signal reconstruction can be ensured in practical applications even in noisy environments by employing a powerful processor with appropriate filtering capabilities.

Table 4.1 Comparison of the execution time

Number of harmonic components, M	Proposed algorithm (s)	FFT (s)	CWT (s)
5	0.0094	0.0131	2.258
7	0.0121	0.0156	2.287
8	0.0128	0.0162	2.321
9	0.0132	0.0165	2.458
11	0.0141	0.0169	2.896
15	0.0185	0.0232	3.125

4.3 Simulation Results

To validate the proposed estimation algorithm, a series of simulation experiments was conducted. The statistical properties of the estimators were assessed using computer-generated noisy samples in accordance with the proposed concept. These samples were obtained by introducing white noise to the samples of the processed signal. In the estimation of deterministic parameters, a widely employed benchmark for evaluating performance is the mean squared error (MSE), with the Cramer-Rao lower bound (CRLB) often serving as a lower bound for the achievable MSE [3, 30–32]. Figure 4.2 illustrates the MSE values for the first five amplitudes of the processed signal, averaged over 10^5 simulations. From the obtained results, it is evident that the proposed estimation procedure approaches the CRLB asymptotically, as demonstrated in [31, 33].

During the simulation process, a sinusoidal test signal with a frequency of 50 Hz was utilized as the input signal. This signal was corrupted by additive white Gaussian noise with a centered distribution, in order to achieve the desired signal-to-noise ratio (SNR). Figure 4.3 presents the maximum errors observed in the estimation of harmonic magnitudes, using input signals of frequencies 30, 50, and 70 Hz, with SNR values ranging from 40 to 70 dB [3]. It is worth noting that in practical scenarios, the SNR of voltage signals obtained from power systems typically falls within the range of 50–70 dB.

To gain further understanding of the performance of the proposed estimation method, we conducted simulations using Matlab and SIMULINK software. The simulations focused on the operation of a standard sigma-delta ADC with an effective resolution of 24 bits and a sampling rate of $f_S = 1$ kHz. The input signal used in

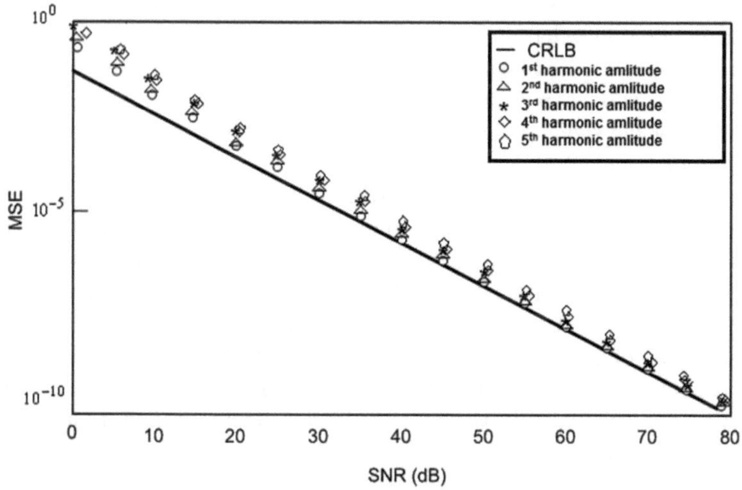

Fig. 4.2 MSE of the five harmonic amplitudes as a function of SNR

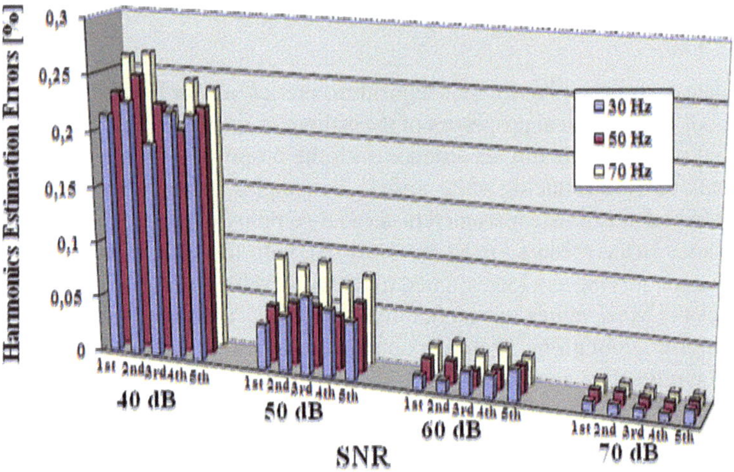

Fig. 4.3 Error of estimation of amplitudes for the first five harmonics, for varying SNR values

these simulations had a fundamental frequency of $f = 50$ Hz.

$$\begin{aligned} x(t) = {} & 1\sin(2\pi f t + \pi) + 0.81\sin(4\pi f t + \pi/3) + 0.62\sin(6\pi f t) \\ & + 0.58\sin(8\pi f t + \pi/6) + 0.41\sin(10\pi f t + \pi/4) \\ & + 0.33\sin(12\pi f t + \pi/12) + 0.16\sin(14\pi f t) \end{aligned} \tag{4.31}$$

The Power Spectral Density (PSD) of the ideal thermal noise and clock jitter affecting the input signal ranged from −100 to −170 dB, resulting in a signal-to-noise distortion ratio (SNDR) between 60 and 80 dB. In the simulation, the combined effect of noise and jitter introduced a relative error of 0.01% in the detection of the fundamental frequency. Figures 4.4 and 4.5 provide a comparison of the simulation results, specifically the maximum errors observed in the estimation of harmonics magnitude and phase, for the signal defined in Eq. (4.31), using the proposed reconstruction algorithm, FFT, and CWT.

A further comparison between the previously mentioned reconstruction procedures and the method proposed in this study was performed using the power spectral density (PSD) of a 1-s section of the processed signal (Eq. 4.31), as shown in Fig. 4.6. To mitigate the leakage effect in the FFT, the Hanning window was applied.

The results obtained from the simulations indicate that the proposed algorithm achieves a level of accuracy comparable to that reported in previous studies [10, 31, 34], and outperforms the method presented in [27]. The simulations confirm that the proposed technique delivers precise estimates, demonstrating its suitability for real-time applications.

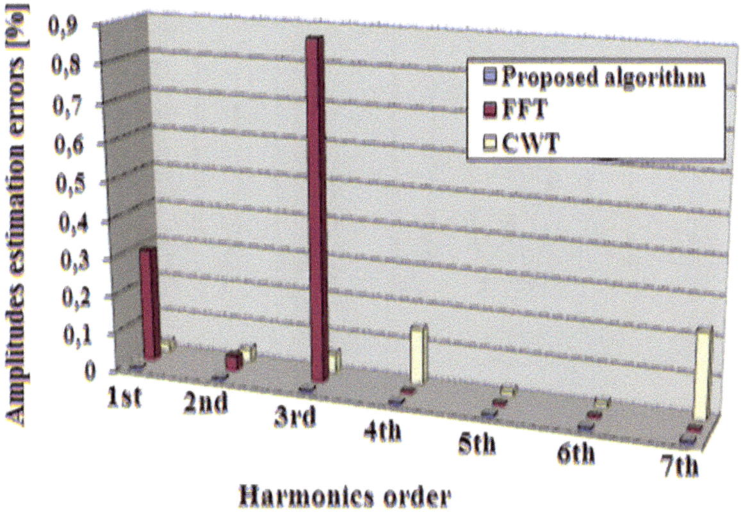

Fig. 4.4 Comparison of estimation errors observed in harmonics magnitude

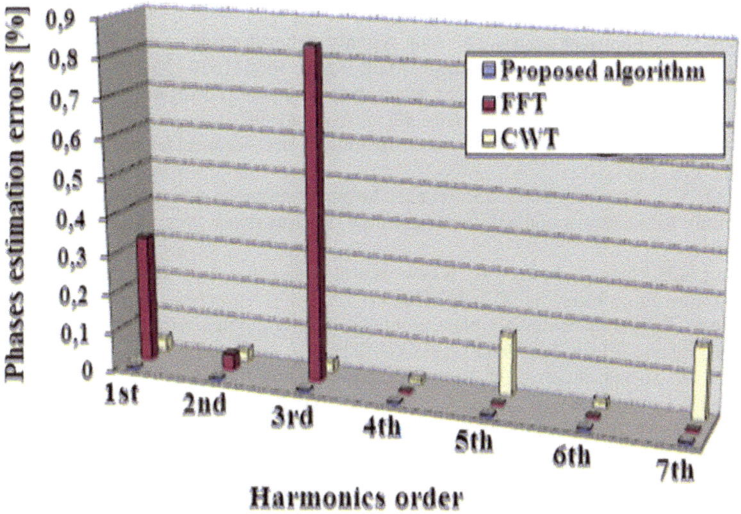

Fig. 4.5 Comparison of estimation errors observed in harmonics phase

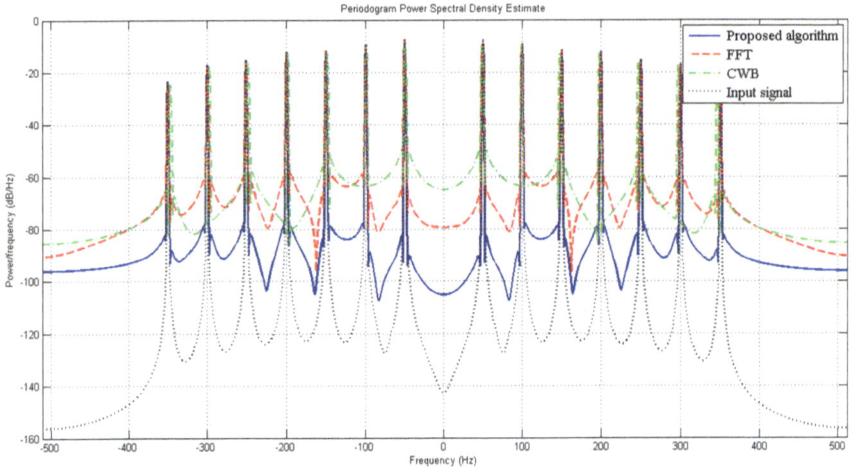

Fig. 4.6 The PSD of the processed signal (4.31) for the various estimation methods

References

1. Terzija, V.V.: Improved recursive Newton-type algorithms for frequency and spectra estimation in power systems. IEEE Trans. Instrum. Meas. **52**(5), 1654–1659 (2003)
2. Sidhu, T.T.: Accurate measurement of power system frequency using a digital signal processing technique. IEEE Trans. Instrum. Meas. **48**(1), 75–81 (1999)
3. Petrovic, P.: Frequency and parameter estimation of multi-sinusoidal signal. Meas. Sci. Rev. **12**(5), 175–183 (2012)
4. Huang, N.E., Shen, Z., Long, S.R., et al.: The empirical mode decomposition and the Hubert spectrum for nonlinear and non-stationary time series analysis. Proc. R. Soc. A **454**, 903–995 (1998)
5. Huang, N.E., Shen, Z., Long, S.R.: A new view of nonlinear water waves: the Hilbert spectrum. Annu. Rev. Fluid Mech. **31**, 417–457 (1999)
6. Petrovic, P., Damljanovic, N.: New procedure for harmonics estimation based on Hilbert transformation. Electr. Eng. **99**(1), 313–323 (2017)
7. Gabor, D.: Theory of communication. J. Inst. Electr. Eng. **93**, 429–457 (1946)
8. Gohberg, I., Olshevsky, V.: The fast generalized Parker-Traub algorithm for inversion of Van der Monde and related matrices. J. Complex. **13**(2), 208–234 (1997)
9. Neagoe, V.E.: Inversion of the Van der Monde matrix. IEEE Signal Process. Lett. **3**(4), 119–120 (1996)
10. Petrovic, P., Stevanovic, M.: Algorithm for Fourier coefficient estimation. IET Signal Proc. **5**(2), 138–149 (2001)
11. Wang, H., Liu, Y., Zhu, B., Song, Q.: Multiple harmonics fitting algorithms applied to periodic signals based on Hilbert-Huang transform. J. Sens. **2013**, 580152 (2013)
12. Kušljević, M.D.: A simple recursive algorithm for frequency estimation. IEEE Trans. Instrum. Meas. **53**(2), 335–340 (2004)
13. Routray, A., Pradhan, A.K., Rao, K.P.: A novel Kalman filter for frequency estimation of distorted signals in power systems. IEEE Trans. Instrum. Meas. **51**(3), 469–479 (2002)
14. Yang, J.Y., Yu, C.S., Liu, C.W.: A new method for power signal harmonic analysis. IEEE Trans. Power Del. **20**(2), 1235–1239 (2005)
15. López, A., Montaño, J.C., Castilla, M., Gutiérrez, J., Borrás, M.D., Bravo, J.C.: Power system frequency measurement under nonstationary situations. IEEE Trans. Power Del. **23**(2), 562–567 (2008)

16. Tomić, J.J., Kušljević, M.D., Vujičić, V.V.: A new power system digital harmonic analyzer. IEEE Trans. Power Del. **22**(2), 772–780 (2007)
17. Nguyen, T.T., Li, X.J.: Application of a z-transform signal model and median filtering for power system frequency and phasor measurements. IET Gener. Transm. Distrib. **1**(1), 72–79 (2007)
18. Poberezhskiy, Y.S., Poberezhskiy, G.Y.: Sampling and signal reconstruction circuits performing internal antialiasing filtering and their influence on the design of digital receivers and transmitters. IEEE Trans. Circuits Syst.-I **51**(1), 118–129 (2004)
19. Romero, D.E.T., Dolecek, G.J.: Digital FIR Hilbert Transformers: Fundamentals and Efficient Design Methods, chap. 19, pp. 451–465. INTECH Open Access Publisher (2012)
20. Oppenheim, A.V., Schafer, R.W.: Discrete-Time Signal Processing. Prentice Hall, USA (1989)
21. Tai, Y.L., Lin, T.P.: Design of Hilbert transformers by multiple use of same subfilter. Electron. Lett. **25**(19), 1288–1290 (1989)
22. Lim, Y.C., Yu, Y., Saramaki, T.: Optimum masking levels and coefficient sparseness for Hilbert transformers and half-band filters designed using the frequency-response masking technique. IEEE Trans. Circuits Syst.-I: Reg. Pap. **52**(11), 2444–2453 (2005)
23. Nigham, N.J.: *Accuracy and Stability of Numerical Algorithms*, 2nd edn. SIAM (2002)
24. Arpaia, P., Serra, C., Daponte, A., Monteiro, C.L.P.: A critical note to IEEE 1057-94 standard on hysteretic ADC dynamic testing. IEEE Trans. Instrum. Meas. **50**(4), 941–948 (2001)
25. Cooklev, T.: An efficient architecture for orthogonal wavlet transforms. IEEE Signal Proc. Lett. **13**(2), 77–79 (2006)
26. Schoukens, J., Rolain, Y., Simon, G., Pintelon, R.: Fully automated spectral analysis of periodic signals. IEEE Trans. Instrum. Meas. **52**(4), 1021–1024 (2003)
27. Tse, N.C.F., Lai, L.L.: Wavelet-based algorithm for signal analysis. EURASIP J. Adv. Signal Process. **2007**, 38916 (2007)
28. Bussow, R.: An algorithm for the continuous Morlet wavelet transform. Mech. Syst. Signal Process. **21**(8), 2970–2979 (2007)
29. Reeves, S.J.: An efficient implementation of the backward greedy algorithm for sparse signal reconstruction. IEEE Signal Proc. Lett. **6**(10), 266–268 (1999)
30. Stoica, P., Li, H., Lim, J.: Amplitude estimation of sinusoidal signals: survey, new results, and an application. IEEE Trans. Signal Process. **48**(2), 338–352 (2000)
31. Belega, D., Dallet, D., Slepicka, D.: Accurate amplitude estimation of harmonic components of incoherently sampled signals in the frequency domain. IEEE Trans. Instrum. Meas. **59**(5), 1158–1166 (2010)
32. Kay, S.M.: Modern Spectral Estimation: Theory and Applications. Prentice-Hall, USA, Englewood Cliffs, NJ (1988)
33. Pantazis, Y., Roces, O., Stylianou, Y.: Iterative estimation of sinusoidal signal parameters. IEEE Signal Process. Lett. **17**(5), 461–464 (2010)
34. Agrež, D.: Improving phase estimation with leakage minimisation. IEEE Trans. Instrum. Meas. **54**(4), 1347–1353 (2005)

Chapter 5
Computational Effective Modified Newton–Raphson Algorithm for Power Harmonics Parameters Estimation

Accurate estimation of harmonics plays a crucial role in mitigating the adverse effects of harmonic components, minimizing losses, and ensuring the delivery of high-quality current. While the FFT algorithm has been widely used, its limitations have led to the development of various other algorithms in the past two decades, broadly categorized as parametric and non-parametric methods [1, 2]. Non-parametric techniques include wavelet transform, Hilbert-Huang transform, Chirp z-transform, and FFT, whereas parametric approaches involve Kalman filters, artificial neural networks (ANN), and Adaline [3–7].

In the context of modern energy systems, which are characterized by complexity, dynamics, and high noise levels, efficient monitoring and evaluation of harmonic components pose significant challenges. Parametric and recursive algorithms are needed to cope with such noisy systems. For instance, the use of Kalman filter models in [3, 4] enables tracking time-varying parameters in the presence of noise, without prior knowledge of the process and noise characteristics. ANN-based methods, such as BPN and RBFNN, have also been proposed [5, 6]. However, these approaches require large amounts of training data, which can be difficult to obtain due to the dynamic nature of power systems and the time-varying characteristics of nonlinear loads. Real-time adaptation of ANN layer structures is another challenge [2, 6]. Adaline, a simple linear adaptive neural network, has been introduced in [7] for online monitoring of time-varying harmonic components. However, it is sensitive to harmonics that are not included in the Adaline model [2].

Recursive Newton-type algorithms have been proposed in [8, 9], assuming that the system frequency is an unknown parameter. These algorithms simultaneously estimate the spectrum and frequency of the power signal, addressing the issue of sensitivity to frequency variations. The proposed approach involves nonlinear estimation and utilizes a recursive strategy with a forgetting factor, leading to improved convergence and accuracy [8, 9].

Another algorithm for harmonic amplitude and phase estimation is presented in [10], based on a modified Newton–Raphson procedure. To reduce computational

P. Petrović, *Processing, Estimation and Measurement of Signals Parameters in Public Distribution Networks*, Studies in Systems, Decision and Control 502,
https://doi.org/10.1007/978-3-031-43107-4_5

complexity, a recursive estimation algorithm is developed using an enhanced formulation of the Jacobian matrix. The proposed method eliminates the need for calculating the inverse of the Jacobian matrix, thereby achieving faster convergence. Additionally, an improved zero-crossing method and digital filtering technique are employed for initial parameter estimation and supervision of the Newton–Raphson updating steps. Simulation results demonstrate the superior accuracy and robustness of the proposed method in harmonic measurement. It offers significant computational efficiency advantages over standard reconstruction algorithms while maintaining lower numerical error.

Assuming that the input signal has a fundamental frequency of f and is band-limited to the first M harmonic components, it can be expressed as the sum of Fourier components.

$$x(t) = X_0 e^{-t/\tau} + \sum_{i=1}^{M} X_i \sin(i\omega t + \phi_i) + v(t) \tag{5.1}$$

In the given model, where $\omega = 2\pi f$ represents the angular frequency in radians per second, X_0 denotes the magnitude of the decaying dc offset, and τ represents the time constant of the decaying dc offset. Furthermore, X_i corresponds to the amplitude value of the ith harmonic, ϕ_i represents the phase angle of the ith harmonic in radians, M denotes the number of harmonic components in the input signal, $v(t)$ represents the noise, and t indicates time in seconds. The proposed model assumes that the dc component remains constant within the interval during which parameter estimation is performed. By discretizing the signal (5.1), the discrete-time measurement $x[k]$ can be obtained, consisting of M sinusoids $s[k]$ with additive white noise, expressed as follows:

$$x[k] = X_0 e^{-kT_S/\tau} + \sum_{i=1}^{M} X_i \sin(\theta_i k + \phi_i) + v[k] = s[k] + v[k] \tag{5.2}$$

where $X_i > 0$, $\theta_i = i\omega T_S \in (0, \pi)$, $\phi_i \in [0, 2\pi)$, T_S stands for sampling period, and $v[k]$ is an additive white noise with variance σ^2. Assuming that M is known, the problem here is to estimate X_i, θ_i, ϕ_i from the only available signal $x[k]$.

5.1 The Newton–Raphson Method

The estimation of parameters using the Newton–Raphson method involves minimizing the error between the estimations and the observations in a least square sense. This can be achieved through a generalized least square optimization algorithm, which can be formulated as follows: considering an objective function $\mathbf{f}$: $\mathbf{R}^n \rightarrow \mathbf{R}^m$ $(m \geq n)$ (where m is greater than or equal to n), the objective is to find a parameter vector $\mathbf{x}$ that minimizes the absolute value of $|\mathbf{f}(\mathbf{x})|$, or equivalently, minimizes the function $\mathbf{F}(\mathbf{x})$:

$$\mathrm{F}(\mathbf{x}) = \frac{1}{2}\sum_{i=1}^{m}(f_i(\mathbf{x}))^2 = \frac{1}{2}\mathbf{f}(\mathbf{x})^T\mathbf{f}(\mathbf{x}) \tag{5.3}$$

The method proposed in this paper extends the Newton–Raphson method to handle the estimation of the processed signal (5.1), which involves solving a system of nonlinear equations with multiple variables. This approach approximates the nonlinear problem by linearizing it around an initial operating point and iteratively converging to the final solution. To ensure convergence and avoid diverging solutions, it is essential to have a good initial estimate for the unknown parameters in the system of equations. In the case of multidimensional functions, such as the processed signal (5.1), which has multiple unknown parameters including amplitude, phase, and frequency of harmonic components, the following steps are necessary:

$$\mathrm{F}(\mathbf{X}) = \mathbf{0} \tag{5.4}$$

where:

$$\mathrm{F}(\mathbf{X}) = \begin{bmatrix} f_1(\mathbf{X}) & f_2(\mathbf{X}) & \cdots & f_n(\mathbf{X}) \end{bmatrix}^{\mathrm{T}} \tag{5.5}$$

it is assumed that the estimation item is n-dimensional vector function, and:

$$\mathbf{X} = \begin{bmatrix} x_1 & x_2 & \cdots & x_n \end{bmatrix}^{\mathrm{T}} \tag{5.6}$$

n-dimensional vector of the unknown. The solution formed in this way is a form of vector equation:

$$\mathbf{X}^{(k+1)} = \mathbf{X}^{(k)} - \left[F'\left(\mathbf{X}^{(k)}\right)\right]^{-1}\mathrm{F}\left(\mathbf{X}^{(k)}\right), k = 0, 1, 2\ldots \tag{5.7}$$

where k is the iteration step. The matrix $\mathrm{F}'(\mathbf{X}^{(k)})$ is called a Jacobi matrix of the function $\mathrm{F}(\mathbf{X}) = 0$, which is defined as:

$$F'\left(\mathbf{X}^{(k)}\right) = \left\{\frac{\partial f_i^{(k)}}{\partial x_j^{(k)}}\right\} = \begin{bmatrix} \frac{\partial f_1(\bullet)}{\partial x_1} & \frac{\partial f_1(\bullet)}{\partial x_2} & \cdots & \frac{\partial f_1(\bullet)}{\partial x_n} \\ \frac{\partial f_2(\bullet)}{\partial x_1} & \frac{\partial f_2(\bullet)}{\partial x_2} & \cdots & \frac{\partial f_2(\bullet)}{\partial x_n} \\ \vdots & \vdots & & \vdots \\ \frac{\partial f_n(\bullet)}{\partial x_1} & \frac{\partial f_n(\bullet)}{\partial x_2} & \cdots & \frac{\partial f_n(\bullet)}{\partial x_n} \end{bmatrix}^{(k)} \tag{5.8}$$
$$= J\left(\mathbf{X}^{(k)}\right) = \mathbf{J}^{(k)}$$

The iterative procedure in the initial phase involves linearizing the nonlinear vector function $\mathrm{F}(\mathbf{X}) = 0$ around a solution $\mathbf{X}^{(k)}$:

$$\mathrm{F}(\mathbf{X}) = \mathrm{F}\left(\mathbf{X}^{(k)}\right) + \mathbf{J}^{(k)}\Delta\mathbf{X}^{(k)} = \mathbf{0} \tag{5.9}$$

Subsequently, the obtained solution is corrected:

$$\mathbf{X}^{(k+1)} = \mathbf{X}^{(k)} - \left[\mathbf{J}^{(k)}\right]^{-1}\mathrm{F}\left(\mathbf{X}^{(k)}\right) \tag{5.10}$$

The convergence criteria are evaluated at each iteration step to ensure convergence. To facilitate fast convergence and obtain proper initial parameters, a simple algorithm incorporating digital filtering and the zero-crossing (ZC) method is proposed. Although the estimation of the fundamental frequency using the ZC method may not be completely accurate, it provides a close approximation. By combining these techniques, the Gauss–Newton method can achieve rapid convergence. It is important to note that the proposed algorithm can estimate the observed signal parameters successfully without the additional procedures mentioned above. However, this comes at the expense of slower convergence and lower precision in the estimation of the processed parameters, which will be further discussed in the subsequent section detailing the results of the simulation tests.

5.2 Frequency Estimation

Estimating the frequency of a multi-harmonic signal using a limited number of discrete samples in the presence of noise is a crucial task with both theoretical and practical implications. The zero-crossing method is commonly employed due to its simplicity. However, the accuracy of this method is influenced by various factors such as precise zero-crossing detection, quantization error, harmonics, noise, and signal dynamics.

Following the sampling theorem, the angular difference [11] between two adjacent continuous points (as shown in Fig. 5.1) can be calculated using the following equation:

$$\Delta\theta = \omega T_S = 2\pi f T_S \tag{5.11}$$

Given the sampling time interval T_s, the angular difference between consecutive continuous points can be determined as shown in the equation above. In the case of a periodic signal with k sampled points within one period, the frequency estimation can be performed using the following formula:

$$f = 1/kT_S \tag{5.12}$$

In practical applications, the chosen sampling frequency is often not an integer multiple of the frequency of the measured signal. As a result, the value of N, representing the number of samples per period, is a non-integer.

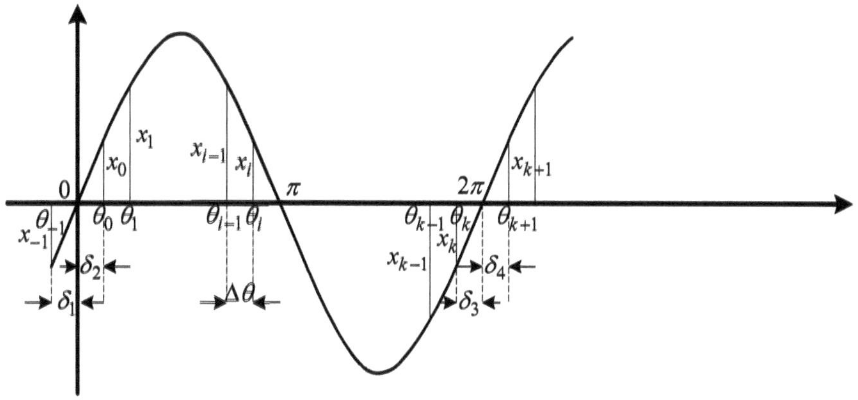

Fig. 5.1 The diagram for zero-crossing detection

Let x_0 and x_k represent the first and last samples of the signal within the observed period. The duration of the period for such a complex harmonic signal can be calculated as follows:

$$T_a = kT_S + \delta_2 + \delta_3 = NT_S = N\Delta\theta \tag{5.13}$$

In the vicinity of zero, the cosine function can be approximated by a linear function, and the value of N can be determined using the following equation:

$$N = k + \frac{\delta_2}{\Delta\theta} + \frac{\delta_3}{\Delta\theta} = k + \frac{\delta_2}{\delta_1 + \delta_2} + \frac{\delta_3}{\delta_3 + \delta_4} \tag{5.14}$$

The angular difference, denoted as $\delta_2 = \theta_0$, represents the angle between the point at which the signal passes through zero and the first sampling point within the observed period. Similarly, $\delta_3 = 2\pi - \theta_k$ denotes the angular difference between the point at which the signal passes through zero and the last sampling point within the observed period. The angular difference between two consecutive samples, denoted as $\Delta\theta = T_s$, corresponds to the sampling period $T_s(=1/f_s)$ (where f_s is the sampling frequency). To determine the number of samples N between 0 and 2π within one period of the processed signal, we use the equation $N = k + \delta_2/\Delta\theta + \delta_3/\Delta\theta$. It is important to note that the sum of δ_2 and δ_3 within a single period may not necessarily match the duration of the sampling period T_s. Consequently, the number of samples N within the period of the multi-sine signal may be a non-integer (fractional number). This modification significantly improves the precision of the frequency estimation algorithm.

Applying the triangle theory [12], we can simplify formula (5.14) presented in Fig. 5.1 to the following form:

$$N = k + \frac{|x_0|}{|x_0| + |x_{-1}|} + \frac{|x_k|}{|x_k| + |x_{k+1}|} \tag{5.15}$$

Hence, a precise estimation of the processed signal can be achieved by utilizing the following expression:

Assuming that more than three cycles of sinusoidal data are being processed, the value of k in Eq. (5.14) can be determined as the number between the first and second maximum within the observed dataset. Subsequently, the first minimum value in the dataset can be identified, and the next $3k$ samples can be stored for further processing. This ensures that the center k samples, ranging from the kth sample to the $(2k + 1)$th sample, represent the observed sine data as depicted in Fig. 5.1. Here, the kth sample corresponds to position θ_{-1}, and the $(2k + 1)$th sample corresponds to position θ_{k+1}. By determining the values of δ_2 and δ_3, as well as obtaining the N value and the frequency f, a precise estimation of the processed signal can be achieved. However, it should be noted that various factors can still affect the accuracy of ZC detection, including higher harmonic amplitudes, non-integer harmonics, and the presence of noise. To address these challenges, a modified Fourier series approach called digital filtering is proposed and implemented in this paper. This approach significantly improves the accuracy of fundamental frequency calculation. Furthermore, additional safety measures have been incorporated into the procedure to identify and discard incorrectly detected ZCs. If the interval between two ZCs significantly deviates from the predetermined interval, the newly detected ZC is deemed invalid, and the earlier result is used instead. This simple logic ensures a favorable level of accuracy in frequency detection of the processed signal, as confirmed through simulation checks conducted using the proposed algorithm.

5.3 Digital Filtering

To enhance the precision of estimating unknown parameters, samples of the processed signal are acquired prior to its utilization as input for the algorithm presented in this paper. These samples undergo an initial transformation using the Fourier non-recursive algorithm [13]. The mathematical formulation of the digital filtering algorithm is expressed as follows:

$$x_i^{fil}(jT_S) = \frac{2}{N}\sum_{k=0}^{N-1} x((k+j)T_S)\cos(ki\omega T),\ j = 1, 2, \ldots, N \tag{5.16}$$

The filtered samples of the ith harmonic voltage or current signal, denoted as $x_i^{fil}(j)$ for the jth sample, are obtained by applying the digital filtering algorithm to the processed signal at each time point $(k + j)T_S$. Here, T_S represents the sampling period, and N represents the number of samples in the initial estimated period. After the filtration process, we obtain a vector of filtered samples with a length of N. This filtered sample vector serves as the input for the proposed procedure, which aims to calculate the unknown parameters of the processed multi-sine signal.

5.4 Proposed Modification of Newton–Raphson Algorithm and Estimation of Harmonics Parameters

The estimation of the unknown dc decaying component of the processed signal is performed using a modified procedure outlined in [14], which reduces numerical errors. By summing N samples of the form (5.2) over the interval corresponding to the period of the processed signal, the sums of the second and third terms approximately cancel each other out, resulting in a value close to zero. However, the sum of the first term, which corresponds to the decaying dc component, remains nonzero. This sum provides an estimation of the unknown dc decaying component of the processed signal.

$$\begin{aligned} \sum_{k=1}^{N} x[k] &\cong \sum_{k=1}^{N} X_0 e^{-kT_s/\tau} = Y_1 \\ \sum_{k=N+1}^{2N} x[k] &\cong \sum_{k=N+1}^{2N} X_0 e^{-kT_s/\tau} = Y_2 \end{aligned} \tag{5.17}$$

The equations presented above allow us to calculate the time constant of the decaying dc component as follows:

$$\tau = T \bigg/ \ln \frac{Y_1}{Y_2} \tag{5.18}$$

By utilizing the Taylor expansion of the exponential function, specifically considering only the first term of the expansion, we can express the sum as follows:

$$Y_3 = \sum_{k=1}^{N} \left(1 + \frac{\ln \frac{Y_2}{Y_1}}{N} \right)^k \tag{5.19}$$

The magnitude of the dc component, which represents the estimation of the constant dc component, can be calculated by taking the mean value of all the samples within a period of the fundamental harmonic component. Mathematically, it can be expressed as:

$$X_0 = Y_1 / Y_3 \tag{5.20}$$

In cases where the time constant of the dc component is low, the estimated value of the dc component can be subtracted from the total input signal samples. This subtraction helps improve the accuracy of the estimation of other harmonic components. However, if the dynamics of the dc component are small, this modification is not necessary, as the proposed algorithm is already capable of achieving high precision in the estimation of unknown parameters.

The proposed method can be categorized as a synchronous method, as the number of samples of the signal being processed, denoted as N, is determined by the ratio of the period of the fundamental harmonic, T, to the sampling period, T_S. It should be noted that N is an integer positive number. Each harmonic component in the processed signal can be represented by the following model, which serves as an initial assumption for the algorithm:

$$x_i(t) = A_i \cos(i\omega t) + B_i \sin(i\omega t),\ i = 1, 2, \ldots, M \tag{5.21}$$

where $\omega = 2\pi/T$. To estimate each componetn of the signal defined with (5.21), it is essential to solve a system of N nonlinear equations, which is overdetermined. The system can be expressed as follows:

$$\begin{aligned} A_i \cos(i\omega T_S) + B_i \sin(i\omega T_S) - x_i^{fil}(T_S) &= f_1(A_i, B_i, i\omega) = 0 \\ A_i \cos(2i\omega T_S) + B_i \sin(2i\omega T_S) - x_i^{fil}(2T_S) &= f_2(A_i, B_i, i\omega) = 0 \\ &\vdots \\ A_i \cos(Ni\omega T_S) + B_i \sin(Ni\omega T_S) - x_i^{fil}(NT_S) &= f_N(A_i, B_i, i\omega) = 0 \end{aligned} \tag{5.22}$$

where $x_i^{fil}(kT_S)$ is defined using relation (5.16). The vector of the unknowns is:

$$\mathbf{X}_i = (A_i, B_i, i\omega)^T \tag{5.23}$$

The Jacobian of the above equation system (5.22) is of the form:

$$\mathbf{J}_i = \begin{pmatrix} \cos(i\omega T_S) & \sin(i\omega T_S) & T_S(B_i \cos(i\omega T_S) - A_i \sin(i\omega T_S)) \\ \cos(2i\omega T_S) & \sin(2i\omega T_S) & 2T_S(B_i \cos(2i\omega T_S) - A_i \sin(2i\omega T_S)) \\ \vdots & \vdots & \vdots \\ \cos(Ni\omega T_S) & \sin(Ni\omega T_S) & NT_S(B_i \cos(Ni\omega T_S) - A_i \sin(Ni\omega T_S)) \end{pmatrix} \tag{5.24}$$

Assume that:

$$F(X_i) = \begin{pmatrix} f_1(A_i, B_i, i\omega) \\ f_2(A_i, B_i, i\omega) \\ \vdots \\ f_N(A_i, B_i, i\omega) \end{pmatrix} \tag{5.25}$$

Upon linearization, the over-determined system of nonlinear equations (5.22) can be transformed into an over-determined system of linear equations in the following form:

$$[\mathbf{F}_i]^{(k)} = -[\mathbf{J}_i]^{(k)}\left(\begin{pmatrix} A_i \\ B_i \\ i\omega \end{pmatrix}^{(k+1)} - \begin{pmatrix} A_i \\ B_i \\ i\omega \end{pmatrix}^{(k)}\right) \quad k = 0,\ 1, 2 \ldots \tag{5.26}$$

In the system (5.26) described above, the initial estimation of the vector of unknown parameters, denoted as $\mathbf{X}_i^{(0)}$, is obtained using the method of least squares. Subsequently, the Newton–Raphson method is employed for further processing based on the proposed concept. The initial values of parameters A_i and B_i are determined as follows:

$$\begin{aligned} A_i^{(0)} &= \frac{2}{N}\sum_{j=1}^{N} x_i^{fil}(j)\cos\left(i \cdot j \cdot \omega^{(o)} T\right) \\ B_i^{(0)} &= \frac{2}{N}\sum_{j=1}^{N} x_i^{fil}(j)\sin\left(i \cdot j \cdot \omega^{(0)} T\right) \end{aligned} \tag{5.27}$$

In the equation above, the value of $\omega^{(0)}$ represents the estimated values of the carrier harmonic frequency, which are obtained using the initial frequency estimation methods such as zero-crossing. By performing a simple transformation of the system (5.26), we arrive at the following equation:

$$\left[\mathbf{J}_i^T \cdot \mathbf{F}_i\right]^{(k)} = -\left[\mathbf{J}_i^T \cdot \mathbf{J}_i\right]^{(k)}\left(\begin{pmatrix} A_i \\ B_i \\ i\omega \end{pmatrix}^{(k+1)} - \begin{pmatrix} A_i \\ B_i \\ i\omega \end{pmatrix}^{(k)}\right) \tag{5.28}$$

Due to the specificities arising in the course of formation of the matrices $[\mathbf{J}_i^T \cdot \mathbf{J}_i]^{(k)}$ and $[\mathbf{J}_i^T \cdot \mathbf{F}_i]^{(k)}$, they can be represented as follows:

$$\left[\mathbf{J}_i^T \cdot \mathbf{J}_i\right]^{(k)} = \begin{pmatrix} a_i^{(k)} & 0 & b_i^{(k)} \\ 0 & a_i^{(k)} & c_i^{(k)} \\ b_i^{(k)} & c_i^{(k)} & d_i^{(k)} \end{pmatrix} \tag{5.29}$$

where:

$$
\begin{aligned}
a_i^{(k)} &= \frac{N^{(k)}}{2};\; b_i^{(k)} = \frac{i \cdot T_S}{4}\left(\left(N^{(k)}+2\right)B_i^{(k)} + A_i^{(k)} ctg(i \cdot x)\right);\\
c_i^{(k)} &= -\frac{i \cdot T_S}{4}\left(B_i^{(k)} ctg(i \cdot x) + N^{(k)} A_i^{(k)}\right);\; x = \frac{2\pi}{N^{(k)}}\\
d_i^{(k)} &= \frac{(i \cdot T_S)^2}{N^{(k)}}\begin{pmatrix}
\left(A_i^{(k)}\right)^2\left(\frac{\left(N^{(k)}+1\right)\left(N^{(k)}-1\right)}{6} - \frac{1}{4}ctg^2(i \cdot x)\right)\\
+\left(B_i^{(k)}\right)^2\left(\frac{\left(N^{(k)}+1\right)\left(N^{(k)}+2\right)}{6} + \frac{1}{4}ctg^2(i \cdot x)\right)\\
+A_i^{(k)} B_i^{(k)} \frac{\left(N^{(k)}\right)^2}{2} ctg(i \cdot x)
\end{pmatrix}
\end{aligned}
\tag{5.30}
$$

where $N^{(k)}$ is the number of samples corresponding to the kth iteration step. By employing similar reasoning, we can establish the following relationship:

$$
\left[\mathbf{J}_i^T \cdot \mathbf{F}_i\right]^{(k)} = \begin{pmatrix} 0 \\ 0 \\ W_i^{(k)} \end{pmatrix}
\tag{5.31}
$$

where:

$$
\begin{aligned}
W_i^{(k)} &= i \cdot T_S \begin{pmatrix}
\frac{A_i^{(k)} B_i^{(k)}}{2} - \frac{\left(\left(B_i^{(k)}\right)^2 - \left(A_i^{(k)}\right)^2\right)}{4} ctg(i \cdot x)\\
-\frac{B_i^{(k)} F_i^{(k)} - A_i^{(k)} G_i^{(k)}}{N^{(k)}}
\end{pmatrix};\\
F_i^{(k)} &= \sum_{j=1}^{N} j \cdot x_i^{fil}(j) \cdot \cos\left(i \cdot j\omega^{(k)} T_S\right);\\
G_i^{(k)} &= \sum_{j=1}^{N} j \cdot x_i^{fil}(j) \cdot \sin\left(i \cdot j\omega^{(k)} T_S\right)
\end{aligned}
\tag{5.32}
$$

During the subsequent analysis of the transposed matrix resulting from the multiplication of the transposed Jacobian matrix (Eq. 5.30), it is observed that only the elements in the third column of the resulting matrix need to be calculated for the iterative algorithm. This leads to a highly efficient procedure, where the computational load is determined by the number of samples involved.

$$
A_i^{(k+1)} = A_i^{(k)} + \frac{b_i^{(k)}}{a_i^{(k)} d_i^{(k)} - \left(b_i^{(k)}\right)^2 - \left(c_i^{(k)}\right)^2} \cdot W_i^{(k)}
\tag{5.33}
$$

$$B_i^{(k+1)} = B_i^{(k)} + \frac{c_i^{(k)}}{a_i^{(k)} d_i^{(k)} - \left(b_i^{(k)}\right)^2 - \left(c_i^{(k)}\right)^2} \cdot W_i^{(k)} \tag{5.34}$$

$$\omega^{(k+1)} = \omega^{(k)} - \frac{a_i^{(k)}}{a_i^{(k)} d_i^{(k)} - \left(b_i^{(k)}\right)^2 - \left(c_i^{(k)}\right)^2} \cdot W_i^{(k)} \tag{5.35}$$

$$N^{(k+1)} = \frac{2\pi}{T_S \omega^{(k+1)}} \tag{5.36}$$

The updated value of the number of samples, $N^{(k+1)}$, is rounded to the nearest integer in the proposed algorithm. Additionally, the period of the signal, $T^{(k)} = N^{(k)} \cdot T_S$, is adjusted in each step to mitigate potential truncation errors. The estimation process continues until a stopping criterion is satisfied, given by the condition:

$$N^{(k+1)} = N^{(k)} \tag{5.37}$$

Then the estimated value of the *i*th harmonic is defined as:

$$g_i^{(k*)}(t) = X_i^{(k*)} \cdot \cos\left(i\omega^{(k*)} t + \phi_i^{(k*)}\right) \tag{5.38}$$

where $k*$ is the value of k for which the condition (5.37) is fulfil, and:

$$\begin{aligned} X_i^{(k*)} &= \sqrt{\left(A_i^{(k*)}\right)^2 + \left(B_i^{(k*)}\right)^2} \\ \phi_i^{(k*)} &= -arctg \frac{B_i^{(k*)}}{A_i^{(k*)}} \end{aligned} \tag{5.39}$$

In the proposed algorithm for estimating the unknown parameters of the processed power signal, an iterative scheme is employed, combining the Newton–Raphson method with the least squares method. Unlike the traditional Newton–Raphson method, which requires recalculation of the Jacobian matrix in each iteration, the proposed modification only necessitates the recalculation of specific elements of the Jacobian matrix. This simplifies both the computation and programming process. The convergence properties of the algorithm have been derived, and it exhibits quadratic convergence rate [15], where the quadratic term, rather than its coefficient, is the primary factor contributing to convergence. Starting with the Taylor expansion of an equation $f(x)$ at a given value, the deviation can be expressed as follows:

$$\Delta x^{(k)} = \frac{f(x) - f\left(x^{(k)}\right)}{f'\left(x^{(k)}\right)} - \frac{f''\left(x^{(k)}\right)}{2f'\left(x^{(k)}\right)} \left(\Delta x^{(k)}\right)^2 \tag{5.40}$$

By neglecting the quadratic and higher order terms in the proposed algorithm, the deviation at the k-th iteration can be expressed using the first term in the equation mentioned above. We can represent it as follows:

$$\begin{aligned} x^{(k+1)} &= x^{(k)} + \Delta' x^{(k)} = x^{(k)} + \frac{f(x) - f\left(x^{(k)}\right)^2}{f'\left(x^{(k)}\right)} \\ &= x^{(k)} + \Delta x^{(k)} + \frac{f''\left(x^{(k)}\right)}{2f'\left(x^{(k)}\right)}\left(\Delta x^{(k)}\right)^2 = x + \frac{f''\left(x^{(k)}\right)}{2f'\left(x^{(k)}\right)}\left(\Delta x^{(k)}\right)^2 \end{aligned} \tag{5.41}$$

It follows:

$$\Delta x^{(k+1)} = \frac{f''\left(x^{(k)}\right)}{2f'\left(x^{(k)}\right)}\left(\Delta x^{(k)}\right)^2 \tag{5.42}$$

The computational burden is a crucial factor in assessing the performance and suitability of spectral estimation techniques for real-time applications. The relative computational burden of various techniques is discussed in [2]. Parametric methods, in particular, tend to have high computational requirements due to the need for eigen decomposition or matrix inversion, which involves operations of the order of m^2n and n^3, respectively, for $n \times m$ and $n \times n$ matrices. To address this issue, we propose a highly efficient implementation that significantly reduces the computational complexity of the Newton–Raphson method from quadratic to linear scale, eliminating the need for matrix inversion. The conventional implementation of the Newton–Raphson method typically has a computational complexity of $O(NM^2)$ for an M-dimensional feature vector and N samples (exhibiting a quadratic relationship with the dimension of the feature vectors) [16]. In contrast, our proposed algorithm for estimating the unknown parameters requires only $O(MN)$ flops, thanks to the efficient implementation that achieves a linear relationship. This reduction in complexity becomes even more remarkable as the dimensionality of the feature vectors increases. Furthermore, the space required for implementing the proposed algorithm is $O(M)$, which is advantageous in terms of memory usage.

5.5 Performance Evaluation

To assess the effectiveness and accuracy of the proposed algorithm for harmonic amplitude and phase angle estimation, a series of simulations were conducted using the MATLAB program package. Multiple tests were carried out to thoroughly evaluate the performance and quality of the algorithm.

The effectiveness of the proposed Zero-Crossing (ZC) technique was evaluated by estimating the fundamental frequency in the presence of harmonics. A synthesized distorted source voltage signal was created, consisting of a fundamental frequency of

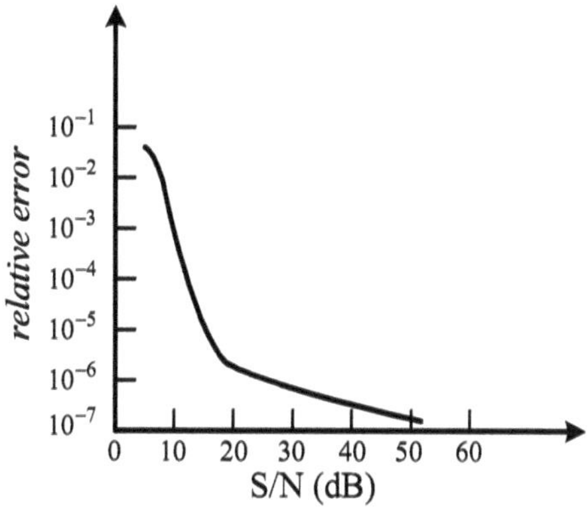

Fig. 5.2 Relative error in estimation of frequency versus input SNR

50 Hz (100%), a 30% third harmonic component, and a 20% fifth harmonic component. The phases of the harmonic components were randomly assigned. Additionally, additive white zero-mean Gaussian noise was introduced to the signal.

Figure 5.2 illustrates the relative error in the estimation of the fundamental harmonic frequency using the proposed ZC technique, as a function of the input signal-to-noise ratio (SNR). The performance of the technique noticeably degrades for SNR values below 10–20 dB. This degradation is attributed to the noise generating its own zero-crossings. As expected, the threshold where performance degrades is lower for high-frequency sinusoids.

To compare the performance of the proposed method with existing methods, various features of the estimated harmonic amplitudes and phases were computed and compared:

- Mean of Absolute Error (MAE): This metric measures the proximity of the estimated amplitudes and phases to their reference values for each harmonic.
- Root Mean Square Error (RMSE): The estimation error variance of each harmonic's amplitude and phase is calculated as the RMSE. This metric reflects the stability of the estimation and is mainly influenced by noise. A lower RMSE indicates higher robustness against noise.
- Mean Squared Error (MSE): The MSE is used as an overall measure of performance and precision for the estimation method. It quantifies the deviation between the estimated waveform and the measured waveform.

By analyzing these metrics, the accuracy, stability, and overall performance of the estimation method can be assessed.

5.6 Static Signal

The performance of the proposed reconstruction algorithm in estimating harmonic amplitudes and phases in the presence of noise was evaluated through a series of simulation tests. The test signal consisted of a fundamental frequency of 50 Hz with higher harmonics (3rd, 5th, 7th, 11th) and a slowly decaying dc component. The signal was generated using MATLAB, as depicted in Fig. 5.3, for various signal-to-noise ratios (SNR).

$$\begin{aligned} x(t) &= 1.5\sin(\omega t + 80^\circ) + 0.5\cos(3\omega t + 60^\circ) \\ &+0.2\sin(5\omega t + 45^\circ) + 0.15\sin(7\omega t + 36^\circ) \\ &+0.1\sin(11\omega t + 30^\circ) + 0.5\exp(-5t) + v(t) \end{aligned} \tag{5.43}$$

This type of signal is commonly observed in industrial loads, such as power-electronic converters and arc furnaces, which are known to introduce extreme conditions in power systems. The random noise component ($v(t) = K\mathrm{rand}(t)$) in the signal follows a normal distribution with zero mean and unity variance. The amplitude factor K represents the variability of the noise, which was varied in this study. The steady-state waveforms obtained using the proposed estimation method exhibited excellent performance, even in the presence of highly polluted and imbalanced grid conditions, as shown in Fig. 5.3.

Table 5.1 provides a comparison of the performance of well-known algorithms [17–19] with the proposed algorithm for estimating harmonic parameters of the fundamental, 3rd, 5th, 7th, and 11th order harmonics. The results clearly demonstrate that the proposed algorithm outperforms other algorithms in terms of errors in amplitude, phase, frequency, and mean squared error (MSE). This algorithm offers powerful tools for accurately tracking steady-state conditions in power networks and power quality monitoring. Furthermore, it was observed that the estimation error in the signal decreases significantly with an increase in the sampling frequency, with the minimum error of 0.0008 achieved at a sampling frequency of 40 kHz.

Figure 5.4 illustrates the Mean Absolute Error (MAE), Mean Squared Error (MSE), and Root Mean Squared Error (RMSE) in the estimation of amplitudes, phases, and frequencies of the signal (5.43) at various signal-to-noise ratios.

Table 5.2 presents the computed Mean Squared Error (MSE) values for the aforementioned algorithms, along with their corresponding computational times.

Upon analyzing the proposed estimation procedure for the unknown parameters of the processed signal, it has been observed that the required number of iterations for convergence is practically unaffected by the Signal-to-Noise Ratio (SNR). Notably, when the sampling period is $T_S = 10^{-4}$ s, convergence is achieved within a mere 3 iterations. Similarly, for $T_S = 5 \times 10^{-5}$ s and $T_S = 2.5 \times 10^{-5}$ s, convergence is attained after 5 iterations. This demonstrates the effectiveness and rapid convergence of the proposed algorithm, rendering it highly suitable for real-time applications. Furthermore, the algorithm exhibits versatility in handling changes in the signal shape, enabling estimation with minimal delay and effectively capturing the dynamics of such signals.

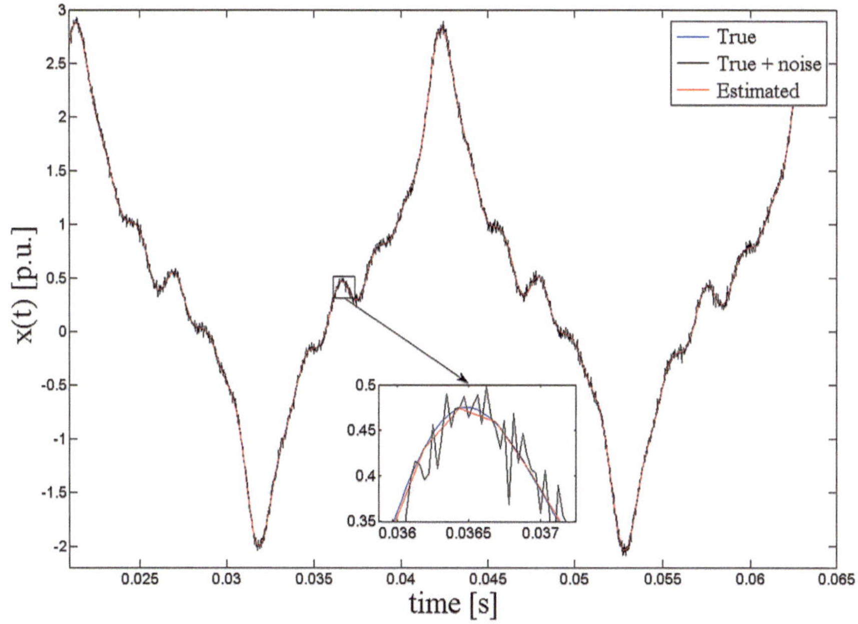

(a)

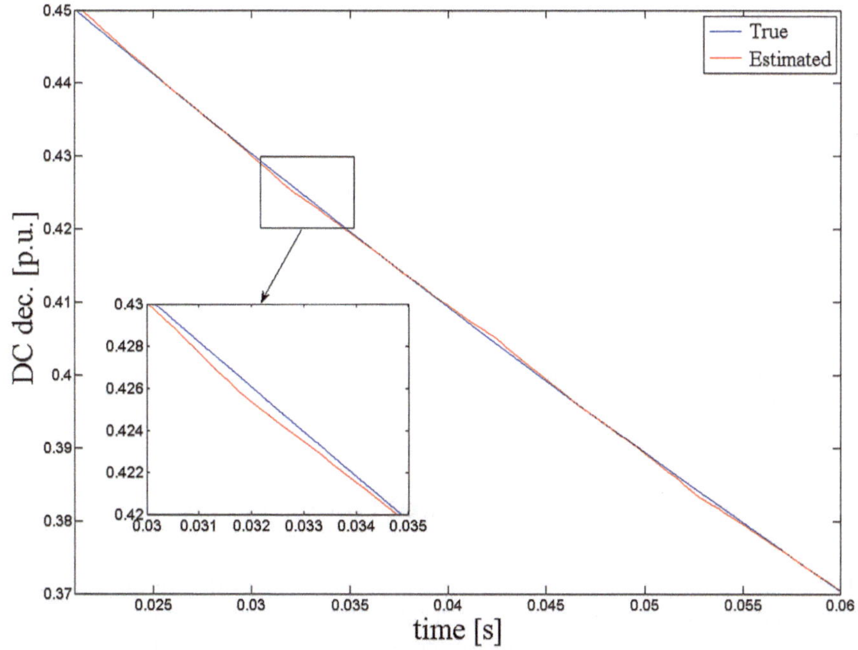

(b)

Fig. 5.3 **a** Harmonic signal defined with Eq. (5.40) (SNR = 30 dB) and estimated waveform, **b** estimated decaying dc component

Table 5.1 Comparison of the methods in the case of processing of static signal (SNR = 60 dB)

Harmonic order		LS-Adaline		Adaptive KF		BRLS*	EnKF	Proposed method		
		MAE	RMSE	MAE	RMSE	MSE	MSE	MAE	RMSE	MSE
1st	Amplit.	0.0005	0.0017×10^{-4}	0.0016	0.0298×10^{-4}	0.012	0.0025	5.317×10^{-8}	2.306×10^{-4}	2.3×10^{-4}
	Phases	0.0581	0.0038	0.0771	0.0117	0.042	0.0048	8.44×10^{-7}	9.187×10^{-4}	9.144×10^{-3}
3rd	Amplit.	0.0031	0.0461×10^{-4}	0.0056	0.1070×10^{-4}	0.016	0.0031	6.499×10^{-9}	8.06×10^{-5}	7.27×10^{-5}
	Phases	0.3961	0.0370	0.6893	0.0850	0.015	0.0028	7.43×10^{-6}	2.725×10^{-3}	2.716×10^{-3}
5th	Amplit.	0.0013	0.0207×10^{-4}	0.0030	0.1060×10^{-4}	0.009	0.0035	1.004×10^{-9}	3.17×10^{-5}	2.96×10^{-5}
	Phases	0.7952	0.0391	1.5885	0.7238	0.012	0.0041	2.08×10^{-5}	4.56×10^{-3}	4.53×10^{-3}
7th	Amplit.	0.0016	0.0264×10^{-4}	0.0036	0.1467×10^{-4}	0.013	0.0037	3.7×10^{-9}	6.08×10^{-5}	5.31×10^{-5}
	Phases	0.7919	0.0145	1.1786	0.2651	0.009	0.0521	4.05×10^{-5}	6.37×10^{-3}	6.34×10^{-3}
11th	Amplit.	0.0021	0.0066×10^{-4}	0.0044	0.0767×10^{-4}	0.012	0.0042	6.398×10^{-9}	7.99×10^{-5}	7.129×10^{-5}
	Phases	0.3207	0.1457	1.0725	1.6579	0.017	0.0215	9.98×10^{-5}	9.99×10^{-3}	9.96×10^{-3}

*Performance of Bilinear Recursive Least Square—BLRS algorithm [18] was estimated without dc decaying component;
EnKF-Ensemble Kalman filter [14]

5.7 Harmonic Tracking in Faulted Power Systems

To evaluate the tracking performance of the proposed algorithm in the context of time-varying harmonic components in a faulty transmission system, a simulation was conducted involving a single-line-to-ground fault on one phase of a transmission line, as depicted in Fig. 5.5 [17]. The estimation results obtained from the proposed algorithm were compared against the LS-Adaline and adaptive KF approach [17], which were optimally tuned by selecting appropriate covariance and noise variance

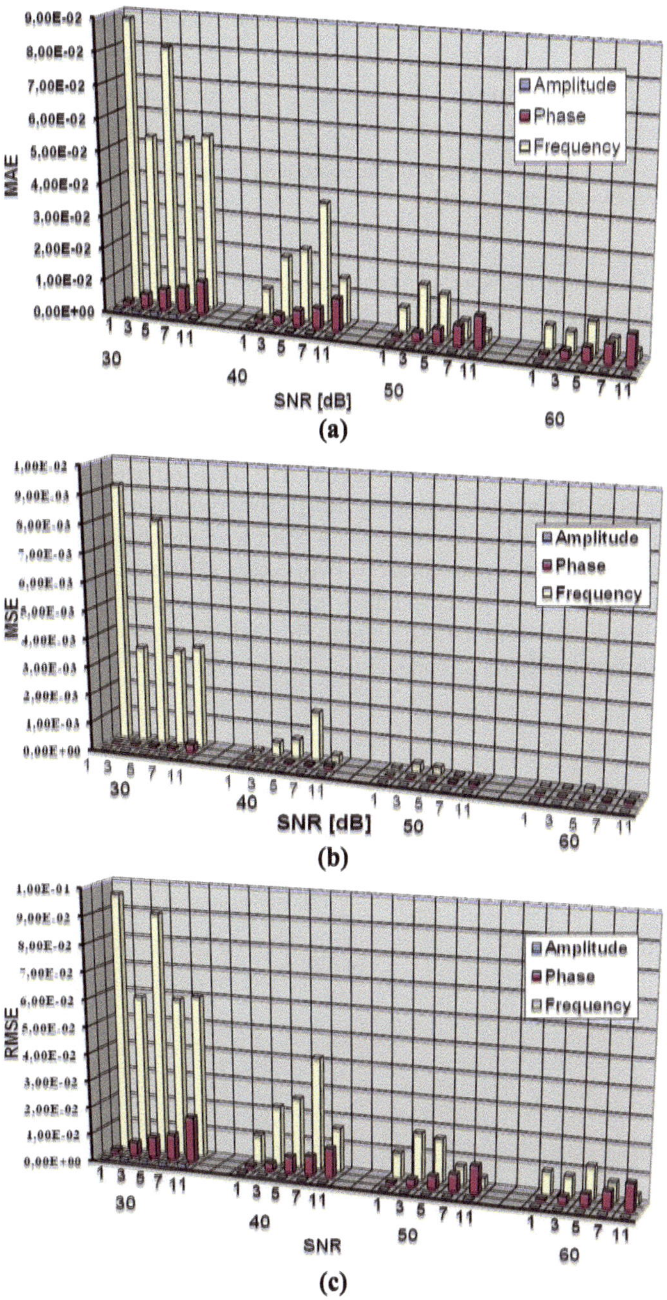

Fig. 5.4 **a** MAE errors as a function of SNR, **b** MSE errors as a function of SNR, **c** RMSE errors as a function of SNR

Table 5.2 Performances of LS-Adaline, Adaptive KF, BLRS, EnKF and proposed algorithm

	LS-Adaline*	Adaptive KF*	BRLS	EnKF	Proposed method
MSE	0.0004	0.0026	0.015	0.0038	3.0678×10^{-6}
CT-computation time (s)	0.212×10^{-3}	0.392×10^{-3}	0.0853	0.086	0.062

*Performance was evaluated after the procedure becoming to track the changes in processed signal (about two periods of observed signal)

values [4]. The comparative results are presented in Table 5.3. Notably, all methods exhibit similar tracking behavior for the harmonic components, and thus, only the tracking of the 3rd harmonic parameter is illustrated in Figs. 5.6 and 5.7. Due to the slow changes in the signal, the estimator demonstrates the ability to track the variations with minimal delay.

To estimate this type of signal, a sliding window approach was employed to capture the dynamics of the observed signal. The time-frame of processed samples was chosen as N/10 samples, providing satisfactory precision (Table 5.3).

The proposed methods were observed to start tracking the abrupt change in the voltage waveform after approximately two cycles (20 ms). However, Table 5.3 demonstrates that the proposed method achieves this tracking with fewer errors and shows less variation in the estimation. This indicates a stronger noise rejection capability, as previously mentioned.

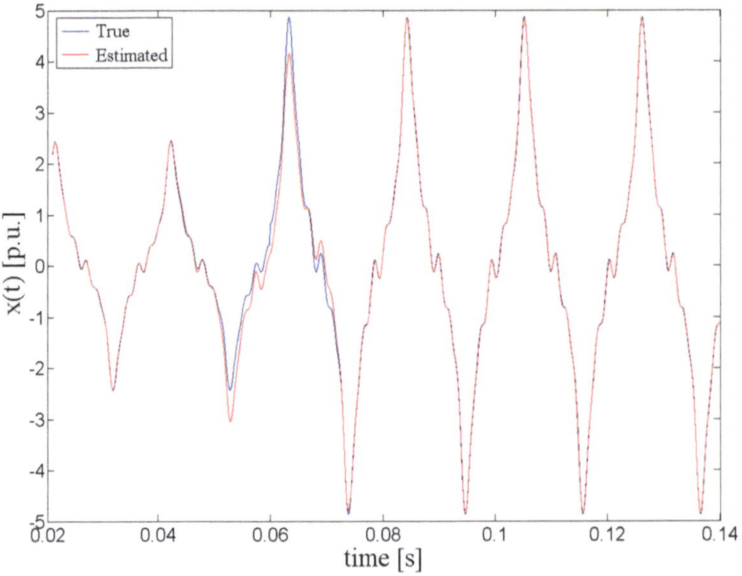

Fig. 5.5 Measured and estimated waveforms in case of fault on the one-phase of a transmission line

Table 5.3 Comparison of the methods in case of faulted power systems

Harmonic order		LS-Adaline		Adaptive KF		Proposed method*	
		MAE	RMSE	MAE	RMSE	MAE	RMSE
1st	Amplit.	0.0011	0.0012×10^{-4}	0.0029	0.0328×10^{-4}	4.6×10^{-4}	4.61×10^{-4}
	Phases	0.0058	0.0001	0.0113	0.0002	9.14×10^{-4}	9.18×10^{-4}
3rd	Amplit.	0.0026	0.0408×10^{-4}	0.0039	0.1606×10^{-4}	1.46×10^{-4}	1.58×10^{-4}
	Phases	0.0350	0.0020	0.0506	0.0027	2.72×10^{-3}	2.73×10^{-3}
5th	Amplit.	0.0033	0.0406×10^{-4}	0.0069	0.3320×10^{-4}	8.45×10^{-5}	1.02×10^{-4}
	Phases	0.1336	0.0057	0.1636	0.0291	4.53×10^{-3}	4.57×10^{-3}
7th	Amplit.	0.0011	0.0101×10^{-4}	0.0013	0.0480×10^{-4}	4.76×10^{-5}	5.94×10^{-5}
	Phases	0.1275	0.0094	0.1623	0.0164	6.33×10^{-3}	6.34×10^{-3}
11th	Amplit.	0.0006	0.0009×10^{-4}	0.0010	0.0181×10^{-4}	3.22×10^{-5}	3.5×10^{-5}
	Phases	0.0969	0.0057	0.2469	0.0392	9.96×10^{-3}	9.96×10^{-3}

*The estimation was performed one period after the change in the shape of the processed signal, while in other methods it was done after the third period

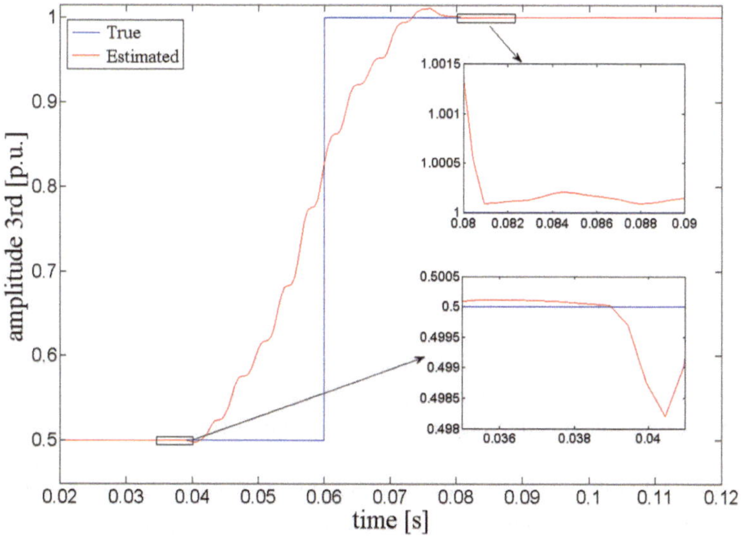

Fig. 5.6 Tracking the 3rd harmonic's amplitude

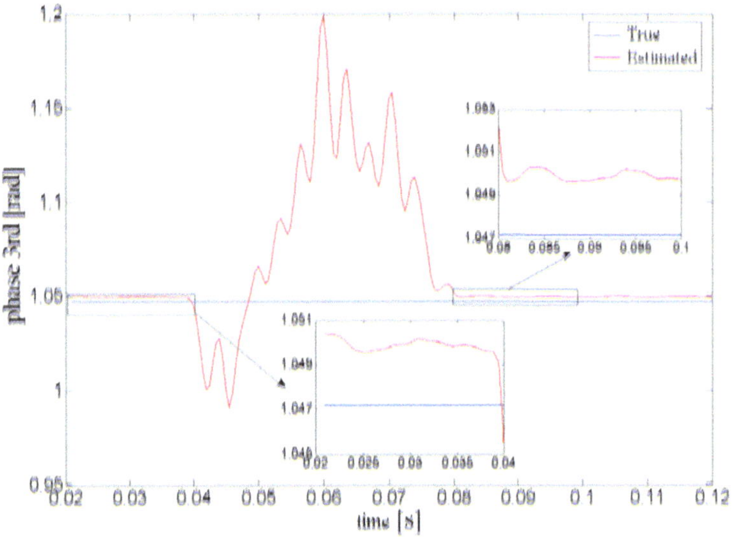

Fig. 5.7 Tracking the 3rd harmonic's phase

5.8 Harmonic Estimation of a Dynamic Signal

In order to assess the tracking capabilities and noise rejection robustness of the proposed algorithm, a time-varying signal in the following form [17] was employed for evaluation:

$$x(t) = [1.5 + a_1(t)]\sin(\omega t + 80°) + [0.5 + a_3(t)]\sin(3\omega t + 60°) \\ +[0.2 + a_5(t)]\sin(5\omega t + 45°) + v(t) \tag{5.44}$$

where:

$$\begin{aligned} a_1(t) &= 0.15\sin(2\pi f_1 t) + 0.05\sin(2\pi f_5 t) \\ a_3(t) &= 0.05\sin(2\pi f_3 t) + 0.02\sin(2\pi f_5 t) \\ a_5(t) &= 0.025\sin(2\pi f_1 t) + 0.005\sin(2\pi f_5 t) \end{aligned} \tag{5.45}$$

and $f_1 = 1$ Hz, $f_3 = 3$ Hz, $f_5 = 6$ Hz. The random noise component, denoted as $v(t)$, follows the same characteristics as the previously defined static signal. The waveform of the processed signal is depicted in Fig. 5.8, while Figs. 5.9 and 5.10 illustrate the tracking of the parameters of the 3rd harmonic using the proposed algorithm. It is worth noting that the estimation of other harmonics follows the same pattern, and thus they are not presented here. To facilitate a thorough comparison of the achieved improvements, performance indices are computed and presented in Table 5.4. These tables provide an assessment of the precision of the proposed algorithm in relation to the LS-Adaline and adaptive KF methods [17]. Notably, the proposed algorithm exhibits a comparable tracking capability to LS-Adaline, initiating the tracking of dynamic harmonic parameters after just one cycle (equivalent to 20 ms). On the other hand, the adaptive KF method requires more than two cycles to achieve the same level of tracking. The results presented in Table 5.4 highlight the superior noise rejection capabilities of the proposed algorithm, particularly for the estimation of the 3rd harmonic.

In comparison to other existing models [4, 17, 20, 21], it has been observed that the proposed solution procedure provides superior accuracy and robustness in estimating the fundamental frequency, amplitudes, and phase angles of all observed frequency components in the measured time-varying signal. The root mean square error (RMSE) performance of the proposed algorithm outperforms the method presented in [22]. Additionally, for both stationary and dynamic signals, the proposed algorithm exhibits better performance compared to the approach described in [23]. The obtained results align with the performance achieved in the processing of static signals, as reported in [9, 24].

The algorithm presented in this study [10] introduces a novel approach for adaptive estimation of amplitudes, frequencies, and phase angles of power harmonics, utilizing the principles of Newton–Raphson method. By introducing a modified Jacobian matrix calculation, the need for computing the inverse matrix in each iteration is eliminated, resulting in simplified programming and reduced processing time. Through simulation experiments, it has been demonstrated that the proposed algorithm outperforms the conventional Newton–Raphson algorithm, providing more reliable estimation of signal parameters and enabling real-time estimation in power systems. Moreover, the computational burden of this algorithm is significantly reduced compared

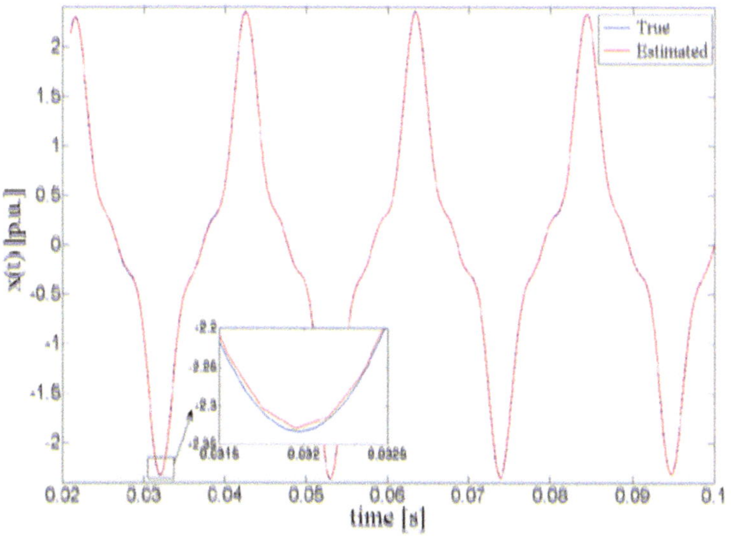

Fig. 5.8 Measured and estimated waveforms in case of time-varying signal (5.44)

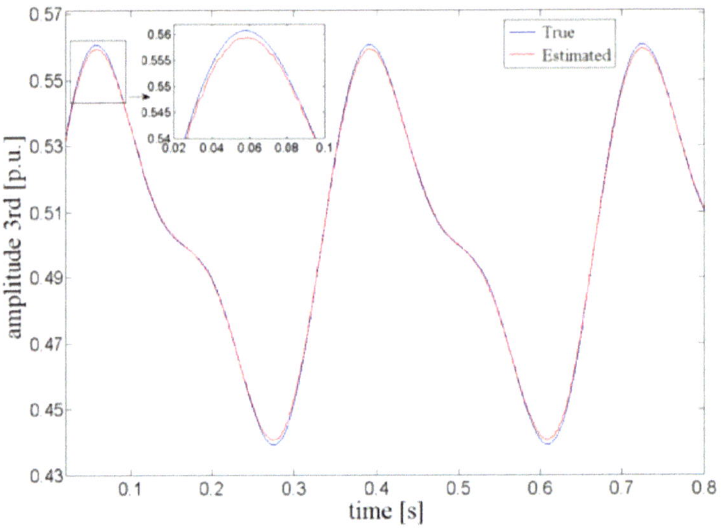

Fig. 5.9 Tracking the 3rd harmonic's amplitude

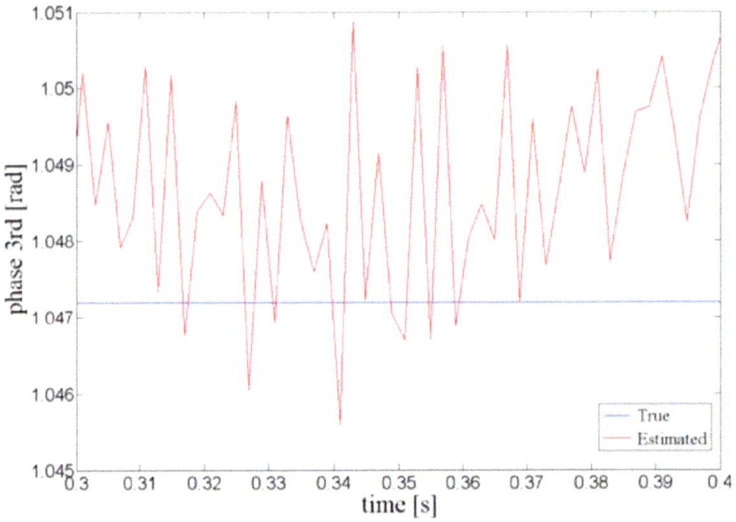

Fig. 5.10 Tracking the 3rd harmonic's phase

to previous methods for frequency and harmonic computations. Therefore, this approach fulfills the time-sensitive requirements of real-time applications in power systems.

Table 5.4 Comparison of the methods for dynamic signal

Harmonic order		LS-Adaline		Adaptive KF		Proposed method	
		MAE	RMSE	MAE	RMSE	MAE	RMSE
1st	Amplit.	0.0112	0.1904×10^{-3}	0.0160	0.4303×10^{-3}	2.21×10^{-3}	2.5×10^{-3}
	Phases	0.0816	0.0125	0.6019	0.5068	1.58×10^{-3}	2.19×10^{-3}
3rd	Amplit.	0.0082	0.1097×10^{-3}	0.0128	0.2628×10^{-3}	1.29×10^{-3}	1.34×10^{-3}
	Phases	0.3875	0.0265	1.5039	3.4814	1.78×10^{-3}	2.05×10^{-3}
5th	Amplit.	0.0067	0.0691×10^{-3}	0.0242	0.8019×10^{-3}	2.1×10^{-4}	2.22×10^{-4}
	Phases	1.1541	1.2591	4.5397	32.0498	4.01×10^{-3}	4.51×10^{-3}

References

1. Alhaj, H.M.M., Mohd, N.M., Asirvadam, V.S., Abdullah, M.F.: Power system harmonics estimation using sliding window based LMS. In: 2013 IEEE International Conference on Signal and Image Processing Applications (ICSIPA), pp. 327–332 (2013)
2. Jain, S.K., Singh, S.N.: Harmonics estimation in emerging power system: key issues and challenges. Electr. Power Syst. Res. **81**(9), 1754–1766 (2011)
3. Macias, J.A.R., Exposito, A.G.: Self-tuning of Kalman filters for harmonic computation. IEEE Trans. Power Deliv. **21**(1), 501–503 (2006)
4. Chen, C.I., Chang, G.W., Hong, R.C., Li, H.M.: Extended real model of Kalman Filter for time-varying harmonics estimation. IEEE Trans. Power Deliv. **25**(1), 17–26 (2010)
5. Wu, X., He, W., Zhang, Z., Dengm, J., Li, B.: The harmonics analysis of power system based on artificial neural network. In: World Automation Congress, pp. 1–4 (2008)
6. Chang, G.W., Chen, C.I., Teng, Y.F.: Radial-basis-function-based neural network for harmonic detection. IEEE Trans. Ind. Electron. **57**(6), 2171–2179 (2010)
7. Sarkar, A., Choudhury, S.R., Sengupta, S.: A self-synchronized ADALINE network for on-line tracking of power system harmonics. Measurement **44**(4), 784–790 (2011)
8. Terzija, V.V.: Improved recursive Newton-type algorithms for frequency and spectra estimation in power systems. IEEE Trans. Instrum. Meas. **52**(5), 1654–1659 (2003)
9. Dash, P.K., Hasan, S.: A fast recursive algorithm for the estimation of frequency, amplitude, and phase of noisy sinusoid. IEEE Trans. Ind. Electron. **58**(10), 4847–4856 (2011)
10. Petrovic, P., Rozgic, D.: Computational effective modified Newton-Raphson algorithm for power harmonics parameters estimation. IET Signal Proc. **12**(5), 590–598 (2018)
11. Vainio, O., Ovaska, S.J.: Authenticating edges produced by zero-crossing algorithms. IEEE Trans. Ind. Electron. **12**, 58–62 (1995)
12. Abu-El-Haija, A.I.: Estimating the parameters of a low frequency sinusoid with noise and harmonic distortion. Int. J. Electron. **62**, 393–410 (1987)
13. Rechka, S., Ngandui, E., Xu, J., Sicard, P.: Performance evaluation of harmonics detection methods applied to harmonics compensation in presence of common power quality problems. Math. Comput. Simul. **63**(3–5), 363–375 (2003)
14. Cho, Y.S., Lee, C.K., Jang, G., Lee, H.J.: An innovative decaying DC component estimation algorithm for digital relaying. IEEE Trans. Power Deliv. **24**(1), 73–78 (2009)
15. Idema, R., Lahaye, D., Vuik, C.: On the convergence of inexact Newton methods. In: Abdulle, A., Deparis, S., Kressner, D., Nobile, F., Picasso, M. (eds.) Numerical Mathematics and

Advanced Applications. Lecture Notes in Computational Science and Engineering, vol. 103, pp. 355–363 (2015)
16. Civek, B.C., Kozat, S.S.: Efficient implementation of Newton-Raphson methods for sequential data prediction. IEEE Trans. Knowl. Data Eng. (2017). arXiv:1701.05378v1
17. Joorabian, M., Mortazavi, S.S., Khayyami, A.A.: Harmonic estimation in a power system using a novel hybrid Least Squares-Adaline algorithm. Electr. Power Syst. Res. **79**(1), 107–116 (2009)
18. Singh, S.K., Goswami, A.K., Sinha, N.: Power system harmonic parameter estimation using Bilinear Recursive Least Square (BRLS) algorithm. Electr. Power Energy Syst. **67**, 1–10 (2015)
19. Pravat, K.R., Bidyadhar, S.: Ensemble-Kalman-filter-based power system harmonic estimation. IEEE Trans. Instrum. Meas. **61**(12), 3216–3224 (2012)
20. Patra, J.P., Dash, P.K.: Fast frequency and harmonic estimation in power systems using a new optimized adaptive filter. Electr. Eng. **95**(2), 171–184 (2013)
21. He, W., Zhaosheng, T., Yong, W., Bo, Z., Xiaoguang, H.: Simple interpolated FFT algorithm based on minimize sidelobe windows for power-harmonic analysis. IEEE Trans. Power Electron. **26**(9), 2570–2579 (2011)
22. Sachin, K.J., Singh, S.N.: Low-order dominant harmonics estimation using adaptive wavelet neural network. IEEE Trans. Ind. Electron. **61**(1), 428–435 (2014)
23. Thirumala, K., Umarikar, A.C., Jain, T.: Estimation of single-phase and three-phase power-quality indices using empirical wavelet transform. IEEE Trans. Power Deliv. **30**(1), 445–454 (2015)
24. Petrovic, P., Stevanovic, M.: Algorithm for Fourier coefficient estimation. IET Signal Proc. **5**(2), 138–149 (2001)

Chapter 6
New CMOS Current-Mode Analogue to Digital Power Converter

Mobile communications terminals or handsets have become ubiquitous in modern society, leading to a growing demand for more efficient power supply solutions for these devices. Accurate monitoring of the transmitting power of mobile terminals plays a crucial role in optimizing power consumption and improving the performance of wireless networks. Power detectors are commonly employed in the power control loop of mobile terminals to detect the transmitting power, enabling power consumption optimization and enhanced linearity of power amplifiers. These detectors are also utilized in gain control loops to optimize the gain of mobile terminals in the transmission or reception chain [1]. RMS power detection is often preferred over peak power detection due to its consistency and standardization in measuring and comparing dynamic signals, regardless of waveform shape.

Various methods have been employed for precision measurement of the power or RMS value of AC voltage, including sampling [2], Monte Carlo [3], and wavelet transform techniques [4]. Traditional power converters, such as full-wave rectifier circuits and multiplier/divider circuits based on the log-antilog principle, have been widely used for this purpose. However, their high-frequency performance is limited to less than 5 MHz due to bandwidth constraints and the slow response of full-wave rectifiers. True RMS-to-DC converters [5] based on bipolar dynamic trans-linear circuits have been proposed, but their implementation requires NPN transistors and operation in only one quadrant, necessitating additional circuits for full-wave rectification. In many cases, MOS technology is preferred for circuit implementation, particularly in mixed A/D systems. One possible solution is the use of sigma-delta converters [6], but the additional required circuits result in a higher transistor count and increased power consumption. Other approaches involve the utilization of up-down translinear loops and class-AB linear transconductors based on CMOS technology [7, 8]. However, these circuits have certain drawbacks, such as operating in only one-quadrant input current, the accuracy-reducing body effect of translinear MOSFETs [9], the need for complex low-pass filters for time-averaging, and restricted input dynamic range due to low voltage operation [7–10].

P. Petrović, *Processing, Estimation and Measurement of Signals Parameters in Public Distribution Networks*, Studies in Systems, Decision and Control 502,
https://doi.org/10.1007/978-3-031-43107-4_6

More recently, current-mode circuits have gained attention for their potential applications, offering wider bandwidth, shorter transition processes, higher slew-rate, increased linearity, wider dynamic range, and lower power consumption. These circuits possess a simpler structure and exhibit improved overall dynamics. It is important to note that processing analog signals within digital circuits introduces inherent challenges, such as jitter errors, occasional data loss, sampling errors, and noise. These limitations often result in insufficient data resolution and high processing errors. The sample and hold circuit, in particular, can contribute to systemic errors in practical implementations.

In light of these considerations, a novel approach to periodic analogue current signal processing is proposed in [11]. The presented converter circuit is primarily designed for determining the mean or effective value of current signals, representing the power or energy of the processed signals. The approach involves integer-based processing of input signals within precisely defined time intervals synchronized with their own period. The core building block consists of an analog current-mode four-quadrant multiplier, integrator, zero-crossing detectors, and a grounded capacitor. The proposed circuit aims to operate at frequencies up to 100 MHz, delivering enhanced linearity and precision in power determination. Unlike existing solutions, this concept enables direct determination of the digital equivalent of the processed value. The HSPICE simulation results corroborate the theoretical analysis, and the analog parts of the proposed circuit have been validated using a reconfigurable structure fabricated in a 0.25 μm CMOS technology.

6.1 Proposed Converter Circuits

Figure 6.1 illustrates the proposed converter circuit. The analogue multiplier takes input current signals of random polarity and performs multiplication. The resulting current product is then fed into the integration circuit, which can be electrically controlled by the bias current I_{B1} (Fig. 6.2). Control over the integration process is achieved through circuits that detect the sign (input duty cycle) of the input current signal, namely the zero-crossing detector 2 (ZCD2) and a T flip-flop triggered by the output of ZCD2. This control scheme involves closing the analogue switch SW1 during every second period of the input current signal being processed. As a result, the integration of the current product is carried out within a time interval equal to the basic period of the input periodic signals, even if they are complex periodic signals.

To ensure that the power calculation in the proposed converter adheres to the definition formula, the integrator's time constant is set to match the period of the input current signals. This is achieved through a detailed procedure described in subsequent sections of this paper.

The output of the T flip-flop also governs the operation of the counter reset. Although not shown in Fig. 6.1, the counter circuit is used to generate the digital equivalent of the calculated power of the input current signals. It is reset at the beginning of each period during which integration takes place. The counter performs

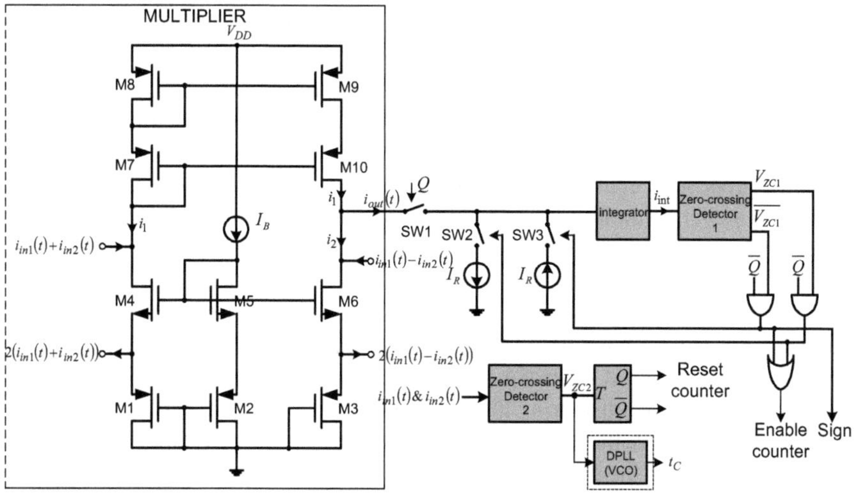

Fig. 6.1 The proposed power converter

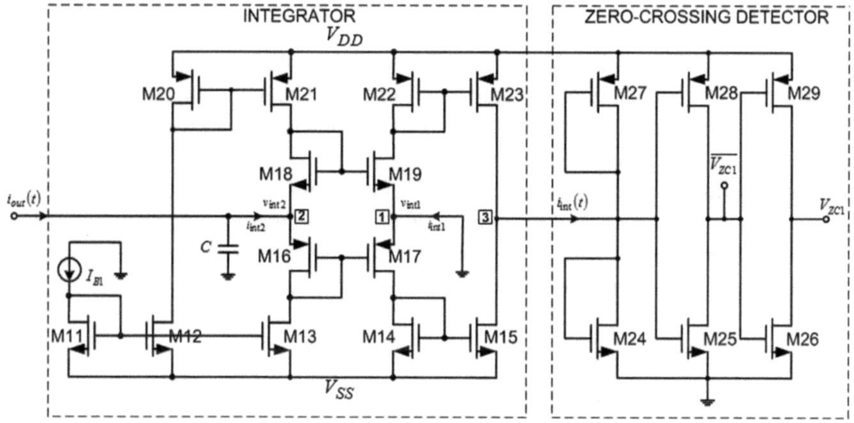

Fig. 6.2 Circuits diagram of integrator and zero-crossing block in proposed architecture

counting within an interval that is a part of and has a maximum value equal to the period following the integration period, regardless of the sign or direction of the current signal at the output of the integrator. The control over the counter's actual counting is established using the output flip-flop signal $\overline{Q}$.

Based on the sign of the output signal from the integrator (detected by ZCD1 in Fig. 6.1), one of the I_R reference current sources is selected through the SW2 and SW3 analogue switches. This selection initiates the discharge process of the integrating capacitor C at a constant current. It practically determines the direction of the current used to discharge the integration capacitor based on the sign of the voltage formed across it.

By measuring the time required for the capacitor to fully discharge (performed by ZCD1), the equivalent of the processed power of the input signals is obtained at the output of the counter. With this processing concept, power calculations for complex periodic current signals can be performed in accordance with the definition formula, enabling the concurrent generation of a digital equivalent of the determined power using the proposed circuit operating in the current mode.

A reset switch, although necessary for the capacitor to initiate circuit operation, is not shown in Fig. 6.1 for the sake of layout simplicity.

The four-quadrant current-mode analogue multiplier depicted in Fig. 6.1 [12] is based on dual translinear loops. The first loop comprises transistors M1, M2, M4, and M5, while the second loop includes transistors M2, M3, M5, and M6. Assuming that all transistors are biased in strong inversion, neglecting channel length modulation and body effect, and considering that the transistors in the two translinear loops are well matched and have the same transconductance value, we can satisfy the condition bn = bp, where b is the physical parameter of the MOS transistor ($\beta_n = \beta_{\mathrm{p}}$) ($\beta = \mu C_{OX} W/L$). This yields the following expression [12]:

$$i_{out}(t) = \frac{i_{in1}(t) i_{in2}(t)}{I_B} \tag{6.1}$$

The DC stability of the output from the current multiplier is maintained even in the absence of feedback or a connected resistor [12]. The input current signals, $i_{in1}(t)$ and $i_{in2}(t)$, can represent signals obtained from voltage and current transducers, or they can be the same signal when calculating the RMS input signal value.

It is important to highlight that in Eq. (6.1), the output current function, $i_{out}(t)$, does not contain any terms related to transconductance parameters or threshold voltages of the MOS transistors, which are sources of temperature effects. As a result, the output current is less susceptible to temperature variations.

6.1.1 Integrator and ZCD Circuits

Figure 6.2 illustrates the implementation of an integrator circuit and zero-crossing detectors. The integrator circuit, as depicted in Fig. 6.2, is designed with a programmable time constant and consists of a single mixed class AB trans-linear loop comprising transistors M16 to M19 serving as the input cell. The utilization of the class-AB technique enables the avoidance of the reduction in high frequency response commonly associated with subthreshold mode operation. To achieve DC biasing (by the current I_{B1}) of the mixed loop, two current mirrors (transistors M11, M12, M20, and M21) are employed, while the input cell exhibits a high impedance input port (for current $i_{out}(t)$) and a low impedance output port (which is grounded). This cell functions as a voltage follower. The output, represented by current $i_{int}(t)$

which replicates the current flow through the grounded port, is realized conventionally using two complementary mirrors. The bias current, I_{B1}, regulates the intrinsic resistance (R_{int}) at the grounded port.

By assuming that $g_{m16} = g_{m18}$ and $g_{m17} = g_{m19}$ we obtain:

$$R_{\text{int}} = \frac{1}{2g_m} = \frac{1}{\sqrt{8\beta_n I_{B1}}} \tag{6.2}$$

where g_{mi} denotes the transconductances of the transistors Mi. Thus, the output current $i_{int}(t)$ can be defined as:

$$i_{\text{int}}(t_{1-2}) = \frac{1}{CR_{\text{int}}} \int_{t_1}^{t_2} i_{out}(t)dt = \frac{\sqrt{8\beta_n I_{B1}}}{C} \int_{t_1}^{t_2} \frac{i_{in1}(t) i_{in2}(t)}{I_B} dt \tag{6.3}$$

The validity of Eq. (6.3) is limited to input currents falling within the ± 10 μA [13] range, as suggested in the proposed converter. When larger currents are present, the intrinsic resistance (R_{int}) becomes nonlinear, rendering Eq. (6.3) invalid. Consequently, the cell's operation is compromised, leading to greater errors.

In the depicted converter circuit (Fig. 6.1), the integrator plays a crucial role in determining its frequency response. The primary restriction on high frequency performance arises from the bandwidth of the negative current mirror M22–M23 and the positive current mirror M14–M15. The dominant poles of these two current mirrors can be approximated as $1/(1 + s(C_{gs22}/g_{m22}))$ and $1/(1 + s(C_{gs14}/g_{m14}))$, respectively, where C_{gsi} denotes the gate-to-source capacitance of MOS transistor Mi.

The zero-crossing detector (ZCD) is implemented using a cascade arrangement of two inverters to generate the switching logic: V_{ZC1} (for $i_{int}(t) > 0$) and $\overline{V}_{ZC1}$ (for $i_{int}(t) < 0$). By detecting the zero-crossing of the processed input current signal with the second ZCD2, edge-triggering of a T flip-flop is performed (the control signal V_{ZC2}), resulting in the generation of a control signal Q at its output (Fig. 6.1).

6.1.2 *Determining the Digital Equivalent of the Processed Power Signal*

Assuming that the time interval between t_2 and t_1 in Eq. (6.3) is equivalent to the input signal period $T = 1/f$, selecting $R_{int}C = T$ allows us to practically determine the average value of the product of input analog current signals based on the defined formula. Consequently, the processed value can be calibrated using the obtained numerical equivalent. In the proposed processing concept, adjustment of the integrator's time constant is achieved by varying the current I_{B1}. However, since this procedure is not automatic, it is necessary to calibrate the detector circuit according to the frequency of the processed current signal.

The performance of the proposed converter is regulated through the utilization of switches SW1, SW2, and SW3, along with the control signals V_{ZC1} and V_{ZC2}. During the interval equal to the T period of the input signal, it is essential to close switch SW1. In the subsequent period, switch SW2 (or SW3 in case when the $i_{int}(t) < 0$) is closed to discharge capacitor C with a constant current I_R.

$$v_C(t) = v_C(t_2) \pm \frac{I_R}{C} t \tag{6.4}$$

The control design described allows for the conversion of bipolar power signals and the generation of the *Sign* signal, with the $\pm$ symbol dependent on the sign of the current $i_{int}(t)$. Specifically, when the voltage $v_C(t)$ is positive ($i_{int}(t) > 0$), ZCD1 generates the control signal V_{ZC1}, which, in conjunction with the signal $\overline{Q}$, facilitates the functioning of the counter.

Assuming that t_3 represents the point at which the capacitor C is fully discharged to zero voltage, we can state the following:

$$v_C(t_3) = v_C(t_2) \pm \frac{I_R}{C}(t_3 - t_2) = 0 \tag{6.5}$$

In accordance with the configuration of the proposed converter, the time interval $t_3 - t_2$ corresponds to the duration during which the counter records N clock impulses with a period of t_c. Therefore, it can be inferred that:

$$t_3 - t_2 = N t_c \tag{6.6}$$

The derived formula indicates that the number N, representing the count registered by the counter at time t_3, is proportional to the average value of the product of the input current signals. This processing approach does not impose limitations on the type of input signals or the number of processed harmonics.

To ensure the automatic adaptability of the proposed converter across different-frequency signals within its entire dynamic range, thus eliminating the need for calibration, Fig. 6.1 introduces an additional circuit, a digital phase-locked loop (DPLL). By employing an internal voltage-controlled oscillator (VCO) at its output, the DPLL generates a clock frequency signal $f_c = 1/t_c$, maintaining a constant ratio between the measured frequency of the processed signal $f = 1/T$ and the generated clock signal. This enables the utilization of a consistent current source I_R throughout the entire bandwidth of the detector. The current value for this source must be determined to ensure the discharge of integration capacitor C within a single period, considering the highest-frequency ($f = 100$ MHz) and largest-amplitude (10 μA) input signal. Furthermore, by adjusting the scaling factor applied to the digital equivalent of the processed input signal's power, in accordance with the established clock frequency f_c, it becomes possible to avoid the need for adjusting the integration constant to match the signal frequency. The scaling factor can be determined based on the known parameters of the DPLL and the counter employed for obtaining the digital equivalent. This approach helps reduce potential processing errors resulting from the non-linear dependence of resistance R_{int} on the applied bias current I_{B1}, while integrating the capacitor C within the IC converter circuit. In such a scenario, these parameters should be set to values that allow the use of a lower current I_R as a necessary component for the discharge of the integration capacitor.

6.2 Non-ideal System Analysis

This section focuses on analyzing the impact of non-idealities in the four-quadrant multiplier and integrator on the performance of the power detector. In Sect. 6.2, the effects of body effect and transistor mismatch in the dual translinear loops of the four-quadrant multiplier were neglected to simplify the circuit calculations. However, considering the body effect, the body-source voltage influences the threshold voltage, causing it to vary as $V_T = V_{T0} + \Delta V_T$ (where V_T is the threshold voltage with substrate bias, V_{T0} is the threshold voltage with zero substrate bias) [12, 14]. To mitigate this effect in N-well technology, PMOS transistors are placed in separate wells where the bulk is connected to the source, ensuring a source-to-body substrate bias of $V_{SB} = 0$ and $V_T = V_{T0}$. In the dual translinear loops, it is possible to place transistors M1, M2, and M3 in separate wells to achieve $V_{SB} = 0$. However, this arrangement is not applicable to transistors M4, M5, and M6, resulting in $V_{SB} \neq 0$. Considering small deviations from the ideal circuit, it can be assumed that $V_{T4} = V_T - \delta$ and $V_{T5} = V_T + \delta$, where $\delta << V_T$. Furthermore, considering the asymmetry of transistors in the dual translinear loops ($k_n \neq k_p$; $k = \beta/2$), the error due to this effect can be calculated by assuming $k_n = k + \Delta k$ and $k_p = k - \Delta k$, where $\Delta k << k$. This approach is more realistic than the example presented in [12], where the non-idealities of the multipliers were considered independently, allowing for a more accurate estimation of errors. These deviations alter the currents $i_1(t)$ and $i_2(t)$ to $i_1'(t)$ and $i_2'(t)$, respectively. By applying Kirchhoff's voltage law (KVL) in the first dual translinear loop and incorporating the above assumptions, we can express the following equations:

$$\sqrt{\frac{i_1'(t) - i_x(t)}{k_p}} + V_T + \sqrt{\frac{i_1'(t) - i_x(t)}{k_n}} + V_T - \delta$$
$$= \sqrt{\frac{I_B}{k_p}} + V_T + \sqrt{\frac{I_B}{k_n}} + V_T + \delta \tag{6.7}$$

where $i_x(t) = i_{in1}(t) + i_{in2}(t)$. After intensive mathematical work and ignoring the terms containing δ^2 and β^2, we obtain:

$$i_1'(t) \approx \frac{i_x^2(t) + 4I_B^2 + 8I_B\delta\sqrt{kI_B} + \frac{4i_x(t)I_B\Delta k}{k} + 8\delta i_x(t)\Delta k\sqrt{\frac{I_B}{k}}}{4I_B + 8\delta\sqrt{kI_B} + \frac{2\Delta k i_x(t)}{k}} \tag{6.8}$$

Applying Kirchhoff's voltage law (KVL) to the second dual translinear loop (for i_2'), we obtain a similar result as Eq. (6.7), but with the substitution of i_x and i_1' with i_y and i_2'. Here, $i_y(t)$ represents the current difference between $i_{in1}(t)$ and $i_{in2}(t)$. Additionally, expressing Kirchhoff's current law (KCL) at the output node leads to:

$$i_{out}'(t) = i_1'(t) - i_2'(t) \approx \frac{i_{in1}(t)i_{in2}(t) + \frac{2i_{in2}(t)I_B\Delta k}{k} + 4\delta i_{in2}(t)\Delta k\sqrt{\frac{I_B}{k}}}{I_B + 2\delta\sqrt{kI_B}} \tag{6.9}$$

By considering the non-ideal integrator characteristics:

$$i_{\text{int2}}(t) = 0;\ v_{\text{int1}}(t) = \alpha v_{\text{int2}}(t) - i_{\text{int1}}(t)R_{\text{int}};\ i_{\text{int}}'(t) = \lambda i_{\text{int1}}(t) \tag{6.10}$$

where $\alpha = 1 - \varepsilon_v$ and ε_v ($|\varepsilon_v| << 1$) represents the voltage tracking error from **2** to **1** terminal, $\lambda = 1 - \varepsilon_c$ and ε_c ($|\varepsilon_c| << 1$) denotes the current tracking error from **1** to **3** terminal of the integrator, respectively. Given the non-idealities, currents generated from integrator can be defined as:

$$\begin{aligned} i_{\text{int}}'(t_{1-2}) &= \frac{\alpha \cdot \lambda}{CR_{\text{int}}}\int_{t_1}^{t_2} i_{out}'(t)dt \\ &\approx \frac{\alpha \cdot \lambda}{CR_{\text{int}}}\int_{t_1}^{t_2} \frac{i_{in1}(t)i_{in2}(t) + \frac{2i_{in2}(t)I_B\Delta k}{k} + 4\delta i_{in2}(t)\Delta k\sqrt{\frac{I_B}{k}}}{I_B + 2\delta\sqrt{kI_B}}dt \end{aligned} \tag{6.11}$$

which results in an absolute error as:

$$Error = \left| i_{\text{int}}(t_{1-2}) - i_{\text{int}}'(t_{1-2}) \right| \tag{6.12}$$

One potential source of error, contributing to nonlinearity in the operation of the proposed integrator, is mobility degradation. To minimize this effect, it is advantageous to maximize the gate-source voltage of the class-AB MOS transistors in the integrator. Another factor that introduces nonlinearity is channel length modulation. This tendency can be mitigated by ensuring a stable drain-source voltage for the class-AB transistors, such as by using a sufficiently large channel length throughout the circuit. Additionally, incorporating cascading into the layout can enhance current mirroring accuracy.

The error expressed in Eq. (6.12) is a function of the input current signals, and its magnitude varies depending on the harmonic components present. This relationship can be deduced from Eq. (6.5), which implies the following:

$$0 \approx \frac{\alpha \cdot \lambda}{C} \int_{t_1}^{t_2} \frac{i_{in1}(t) i_{in2}(t) + \frac{2 i_{in2}(t) I_B \Delta k}{k} + 4\delta i_{in2}(t) \Delta k \sqrt{\frac{I_B}{k}}}{I_B + 2\delta \sqrt{k I_B}} dt \pm \frac{I_R}{C} N t_c \quad (6.13)$$

The precision of the proposed converter is unaffected by the imprecision of the intrinsic resistance R_{int}, which offers a significant advantage in this processing approach. It is worth noting that the obtained resistance exhibits a substantial dependence on process-voltage-temperature (PVT) factors, influenced by the current range and operating temperature of the circuit. However, this resistance does impact the determination of the integrator's output current (Eq. (6.11)), consequently influencing the precision of subsequent processing, particularly through the zero-crossing circuit that defines the counter operation interval. To mitigate the impact of this previously problematic and potentially nonlinear resistance, an external resistor can be added to the implementation of the integrator (at terminal 1).

The extreme PVT variation observed ranges from ±10%. This tolerance is applied across the temperature range of 0–100 °C, considering variations of ±5% in supply voltage and on-die calibrated capacitor. The ±10% tolerance primarily arises from the nonlinearity of bias and reference currents concerning temperature, supply variation, and slight but finite variations in the calibrated capacitor. Statistical simulation results based on Monte Carlo analysis [15] for the proposed converter under typical conditions (supply = ±2.5 V, process = typical, and temperature = 50 °C) demonstrate that the observed 1 s standard deviation for 2000 runs is 0.3 mA, with a mean value of currents generated from the integrator at 3.5 mA. The statistical spread, with a value of $\sigma/\mu = 0.085$, is sufficiently low for the proposed converter to be regarded as highly resilient to potential PVT changes.

6.2.1 Measuring Uncertainty

The measurement uncertainty of the implemented power converter is a critical metrological parameter [16]. To achieve a highly accurate measurement, stringent requirements were imposed, significantly influencing all aspects of the circuit's design, development, and implementation. A comprehensive model can be formulated to represent the measurement results, considering the factors that have a significant impact on the measurement uncertainty of the employed method (derived from Eqs. (6.6) and (6.13)):

$$(N + \Delta N + \delta N) = \frac{\alpha \cdot \lambda}{(I_R + \Delta I_R) t_c} \int_{t_1}^{t_2} \frac{i_{in1}(t) i_{in2}(t) + \frac{2 i_{in2}(t) I_B \Delta k}{k} + 4\delta i_{in2}(t) \Delta k \sqrt{\frac{I_B}{k}}}{I_B + 2\delta \sqrt{k I_B}} dt \quad (6.14)$$

- ΔI_R—the correction of the reference current source is necessary to account for its inherent variability, which follows a random distribution. The mathematical expectation of this variability is zero, indicating an unbiased nature. However, the relative standard measurement uncertainty, denoted as $u_{\Delta R}$, is determined by the manufacturer's specifications and represents the level of uncertainty associated with this type of uncertainty component.
- ΔN—the correction of the N-counter content is essential to account for the precision of the zero comparator's operation. The value ΔN is determined as the ratio of ΔT (the difference between the period of the measured signal and the processed (real) signal) to the clock period t_c. ΔN is a random variable with an even distribution, and its mathematical expectation is zero. The relative standard measurement uncertainty, $u_{\Delta N}$, is of Type B and can be calculated as $u_{\Delta N} = u_{\Delta T}/t_c$, where $u_{\Delta T}$ represents the measurement uncertainty in evaluating the period of the processed signal. In the case where the error of the comparator is eliminated through the application of the "auto-zeroing" procedure, this correction and its associated uncertainty can be neglected.
- δN—the inherent error of the counter is denoted as ± 1, representing a random variable with an even distribution. The mathematical expectation of this error is zero, indicating an absence of bias. The standard measurement uncertainty associated with the counter, denoted as udN, is determined as $u_{\delta N} = 1/(2\sqrt{3})$.

Example In the case of a processed current signal effectively close to the reference current source and utilizing a 16-bit counter operating at a frequency of 10 MHz, along with an error in determining the frequency of the processed signal within ±0.1 Hz, the standard measurement uncertainty of the method is estimated to be approximately 0.35% [16]. An analysis of the measurement uncertainty demonstrates that the evaluation of the processed signal's frequency exerts a dominant influence on the overall uncertainty.

6.3 Simulation Results

To validate the proposed circuit, HSPICE simulations were conducted using TSMC 0.25 μm LEVEL49 (BSIM3v3) CMOS process parameters. The input currents were varied within a significant range from −10 to 10 mA. The constant current in the multiplier IB was set to 10 mA. The circuit was biased with ±2.5 V supply voltages. The input dynamic range of this circuit is determined by the performance capabilities of the multiplier at the input side of the proposed converter and corresponds to $\pm I_B$ [12].

During the simulation process in the proposed converter circuit, well-established solutions were employed for the realization of current sources I_B and I_{B1} through MOS biasing circuits [14]. The reference source I_R was implemented using a DO-CCCII (dual-output current controlled conveyor) [17], where the input port y was connected to a DC voltage source, ensuring stability and adjustability of the current source amplitude. Consequently, the two output ports of the DO-CCCII generated current sources in opposite directions, aligning with the requirements of the proposed converter as depicted in Fig. 6.1. It is worth noting that the realization of such a reference source could also rely on existing solutions from the literature [14, 18, 19]. However, it should be acknowledged that the inclusion of an output-complementary circuit [17, 20] is necessary at the output of the reference source implemented in this manner, due to the requirement for sources with opposite directions.

Figure 6.3a presents plots of the output current signals from the integration circuit for three different input current signals. These plots demonstrate that the output current of the proposed integrator reaches the average power value of the processed input signals at the end of their respective periods. The conversion of this value into a digital equivalent is accomplished by discharging capacitor C with a constant current. The plots are generated for three distinct values of amplitude I_2: (1) $I_2 = 10\ \mu A$; (2) $I_2 = 5\ \mu A$; (3) $I_2 = 1\ \mu A$.

Figure 6.3b illustrates the response of the proposed converter when processing input current signals with a frequency of 100 MHz, comparing it to the ideal scenario response. This allows for an evaluation of the error magnitude within the bandwidth of the proposed converter.

The value of capacitor C is selected based on the frequency of the processed current signals to ensure its discharge within a single period of the processed signal. This capacitor can be a fixed component integrated into the proposed IC circuit. In such cases, the procedure described earlier for determining the digital equivalent of the processed power of input signals with the application of scaling factors would be necessary. However, if the capacitor is detachable from the chip, the proposed solution offers increased flexibility and improved calibration options.

Figure 6.4a presents the DC transfer characteristic of the proposed circuit at different temperatures for the case of $i_{in1}(t) = i_{in2}(t)$. A zoomed-in section near zero is provided as well. The frequency response of the proposed converter is illustrated in Fig. 6.4b. It is noteworthy that the nonlinearity level in the dynamic range remains below 1 dB, resulting in a dynamic range of approximately 20 dB. The detector circuit proposed in this paper exhibits higher linearity compared to the circuits described in [21].

The Crest Factor, which describes the wave shape of a dynamic signal, is commonly defined as the ratio of the peak value to the RMS value of the waveform. For instance, a signal with a crest factor of 4 has a peak value four times higher than its RMS value. The proposed circuit performs exceptionally well for crest factors of 4 or less, while also exhibiting reduced accuracy for signals with higher crest factors (Fig. 6.5). In Fig. 6.5, the term "SCR waveforms" refers to the ideal chopped sine wave. The high linearity achieved throughout the proposed solution directly contributes to its excellent performance with crest factors lower than 4.

The proposed circuit exhibits a high operating frequency capability, reaching up to 100 MHz, while maintaining enhanced linearity and precision in determining power values. The power consumption of the simulated detector was measured at 8.37 mW. Moreover, the circuit introduced in this paper allows for electronic control of the

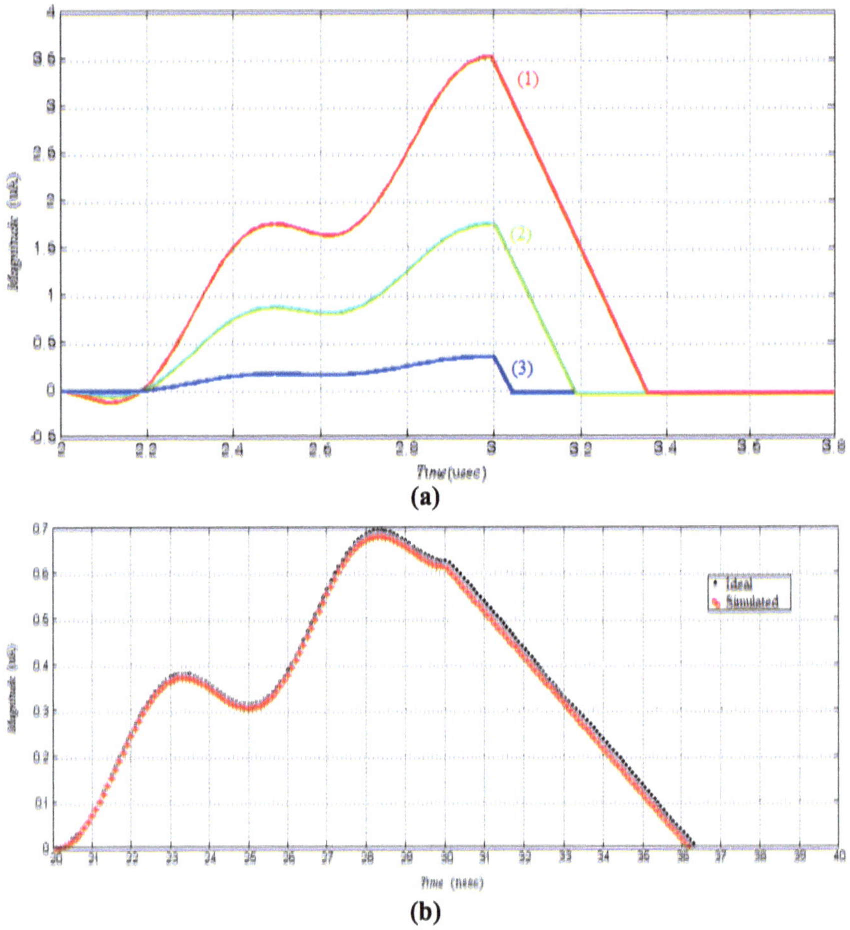

Fig. 6.3 **a** Time-domain response of the proposed power detector for the $i_{in1}(t) = 10\sin(2\pi ft)$; $i_{in2} = I_2\sin(2\pi ft - \pi/4)$, with fundamental frequency $f = 1/T = 1$ MHz. The value of capacitance was selected as $C = 1$nF, while $I_{B1} = 10\,\mu$A ($R_{int} \approx 1$ kΩ) and $I_R = 10\,\mu$A; **b** Time-domain response of the proposed power detector for the i_{in1}(t) $= 5\sin(2\pi ft)$; $i_{in2} = 5\sin(2\pi ft + \pi/3)$, with fundamental frequency $f = 100$ MHz. The value of capacitance was selected as $C = 10$ pF, $I_{B1} = 10\,\mu$A and $I_R = 1\,\mu$A

integrator's time constant, contributing to its adaptability and flexibility. The error in computing the effective value of the processed input current signals was found to be lower compared to previous studies [10, 22].

Figure 6.6 showcases the final layout of the proposed converter, implemented using Cadence Virtuoso XL 6.1.6v software from Texas Instruments. The circuit occupies an area of 68 × 75 μm. The layout design includes a multiplier circuit, two zero-crossing detectors (ZCDs), three analog switches, an integrator, bias currents, and a reference current source (IR) realized using a DO-CCCII. Notably, all the

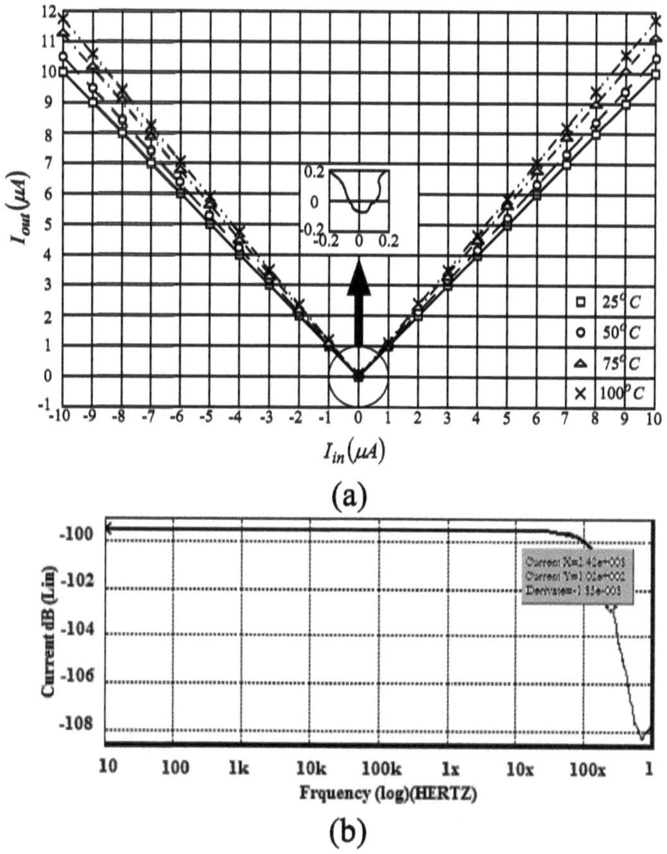

Fig. 6.4 **a** Simulated DC transfer characteristic versus different temperatures, i.e. 25, 50, 75 and 100 °C; **b** simulation result for frequency response of proposed converter

Fig. 6.5 Performance versus crest factor

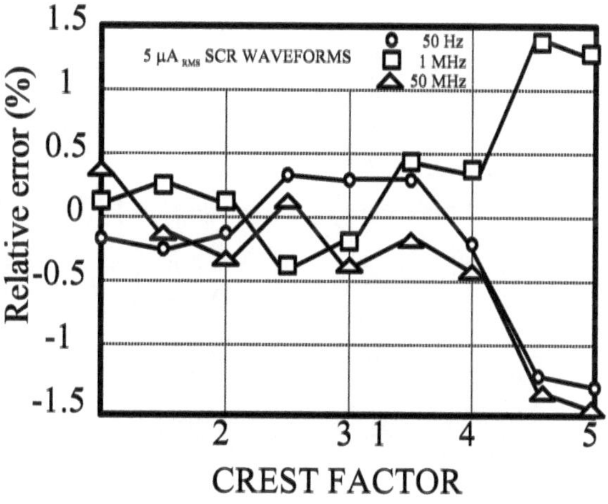

circuit components operate in the current mode, while the capacitor C is detached from the chip as per the proposed solution.

Figure 6.3 displays the results obtained from pure sinusoidal signals. However, for a power (or RMS) detector, it is crucial that the circuit exhibits consistent responses to signals with equal power values but varying waveform shapes. To validate the performance of the proposed converter (analog part) under diverse signal processing conditions, the circuit underwent simulation and experimental testing using various input waveforms. The test input signals and control signals (Q and $\overline{Q}$) were generated using the NI PCI-6259 ELVIS/PCI acquisition card from National Instruments. The output voltages of the card were converted into input currents using external LTP CCII01 current conveyors and 10 kΩ resistors.

Figure 6.7 presents the results of signal processing achieved with the proposed converter, showcasing calibrated signals with different amplitudes, frequencies, and waveform shapes. These include single-tone sinusoids, two-tone signals, square

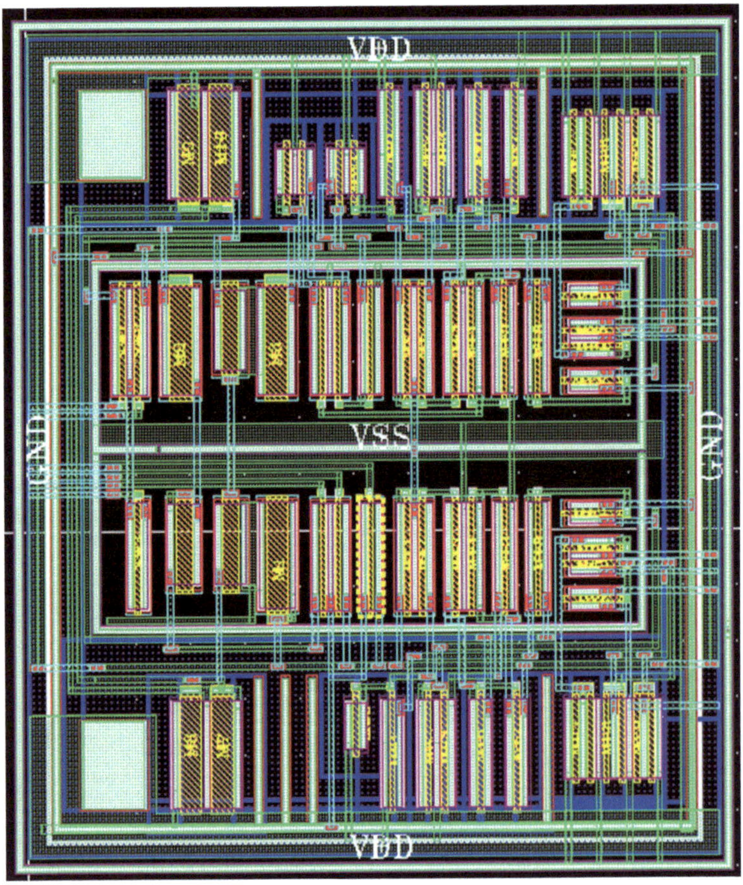

Fig. 6.6 The final layout of proposed converter

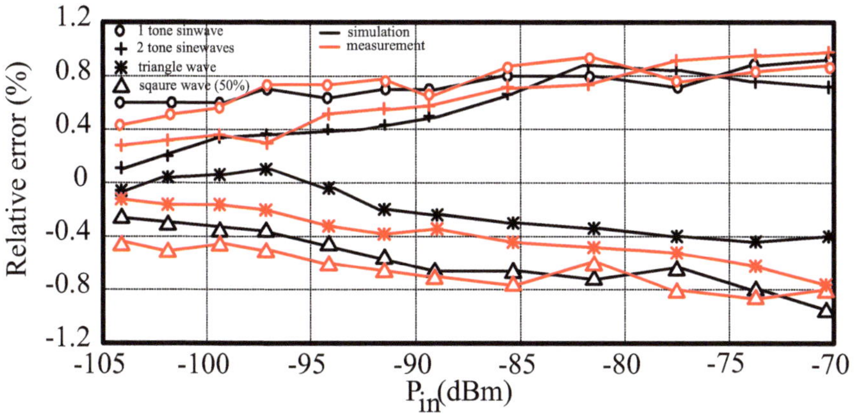

Fig. 6.7 Comparison of the experimental results with simulated data of the proposed converter for various waveforms

waves with a duty cycle of 50%, and triangle waves. All the signals used in the experiments were set at a frequency of 1 MHz. Additionally, parallel measurements of RMS voltage were conducted using a highly precise Fluke 5790A multimeter (with a reference traceable to metrology-grade standards), allowing for the evaluation of the proposed circuit's applicability in real-world scenarios. The relative errors observed were lower than 1% for $P_{in} \leq -90$ dBm. The post-layout simulations and measurement results for the analog part of the proposed converter provided confirmation of the findings obtained during the HSPICE simulation process.

Based on the aforementioned results, it can be concluded that the simulated and experimental outcomes align with the theoretical analysis, thereby validating the performance and accuracy of the proposed circuit design.

6.3.1 Comparison with Existing Circuits

In order to gain a comprehensive understanding of the proposed technique, a comparative analysis was conducted to evaluate the performance of the proposed system in relation to previous implementations of power converters. Table 6.1 provides a summary of key parameters for these converters. It is important to note that not all of the compared implementations are based on the current-mode circuit concept proposed in this paper.

The frequency response and dynamic range of this CMOS power detector were found to be comparable, and in some cases superior, to most diode detectors. The error in computing the effective value of the processed input current signal was lower compared to the majority of the compared implementations. Additionally, the proposed detector circuit, which offers a wider dynamic range, is more practical for

Table 6.1 Comparison of performance of different power converters

Parameter	[7]	[9]	[10]	[22]	[25]	[26]
Basic principle	Class-AB transconductance	SRD translinear	MOS translinear loop	CMOS squaring	Squaring and averaging	Rectification and squaring
Programmability	No	No	No	No	No	No
Technology	0.5 μm	2.4 μm	0.18 μm	0.35 μm	Bipolar	0.35 μm
Verification type	Fabrication	Fabrication	Simulation	Simul. and exper.	Simul. and exper.	Simul. and exper.
Supply voltage	1.5 V	1.5 V	1.2 V	15 V	2–15 V	1.5 V
Circuit complexity (transistor no.)	40	108	18	13	11MOS + 4Bias	Variable
Operation area of input	1 Quad	1 Quad	2 Quad	2 Quad	2 Quad	2 Quad
Power	–	–	<100 μW	6.04 nW	<15 mW	–
Area	0.1 mm^2	0.4 mm^2	–	–		–
Relative error	3%	2%	3%	<2%	0.1–2.5%	>1%
At input range	12–22 μA	5–11 μA	2.5–30 μA	1.5 mA	10 μA–1 mA	500 mV
Maximum input frequency	–	–	–	95 MHz (I_{in} = 1 mA) 66 MHz (I_{in} = 500 μA)	700 kHz–60 MHz	10 MHz
Dynamic range	–		11 dB	12 dB	20 dB	–

(continued)

Table 6.1 (continued)

Parameter	[23]	[27]	[24]	[28]	[29]	[11]
Basic principle	Squaring and filtering	Log-domain FGMOS	MTL cell (SCMC)	SRD FG-MOS	Sqauring and integration	Multiplication and integration
Programmability	No	No	Yes	No	Yes	Yes
Technology	Bipolar	0.35 μm	0.5 μm	0.6 μm	0.35 μm	0.25 μm
Verification type	Fabrication	Simulation	Fabrication	Simulation	Experiment	Simul. and fabric.
Supply voltage	2–16.5 V	0.9 V	2.5–5 V	1.2 V	1.5 V	2.5 V
Circuit complexity (transistor no.)	–	7MOS + 6FG	18	14MOS + 1FG	26MOS + 3OA + digital part	60MOS + digital part
Operation area of input	2 Quad	2 Quad	2 Quad	2 Quad	2 Quad	2 Quad
Power	–	1.5 μW	>88 μW (2 mW)	–	–	8.37 mW
Area	–	–	0.064 mm^2	–	–	0.0051 mm^2
Relative error	0.5–1%	2%	3%	3%	0.05%	0.8%
At input range	200 mV	0.6–400 nA	290–300 μA	5–30 μA	1 V	±10 μA
Maximum input frequency	1 MHz	–	>2 MHz	–	1 MHz	100 MHz
Dynamic range	50 dB	28 dB	–	–	26 dB	20 dB

realization when compared to the solutions described in [23, 24]. Based on these performance metrics, the proposed converter demonstrates its suitability for precise measurements of the RMS value of AC signals in various applications, including instrumentation, display systems, biomedical ICs, automatic gain control systems, and analog spectrum analyzers. Furthermore, it can be utilized in telecommunication and measurement systems for power control of transmitters, multi-tone tests, and electromagnetic interference (EMI) measurements.

The proposed CMOS current-mode power converter introduces a new architecture [11]. By eliminating the need for a separate sample-and-hold circuit, it achieves a digital equivalent of active power, RMS value, or energy for input four-quadrant current signals using definition formulas. This architecture is well-suited for low supply voltage and wide-frequency range applications. It can be easily fabricated in IC form for use in battery-powered or portable electronic devices, such as wireless devices. The HSPICE simulation and experimental results align closely with the anticipated theoretical performance.

References

1. Yijun, Z.: A low-power ultra-wideband CMOS true RMS power detector. IEEE Trans. Microw. Theory Tech. **56**(5), 1052–1058 (2008)
2. Pogliana, U.: Precision measurement of ac voltage below 20 Hz at IEN. IEEE Trans. Instrum. Meas. **46**(2), 369–372 (1997)
3. Germer, H.: High-precision AC measurements using the Monte-Carlo method. IEEE Trans. Instrum. Meas. **50**(2), 457–460 (2001)
4. Novotny, M., Sedlacek, M.: RMS value measurement based on classical and modified digital signal processing algorithms. Measurement **41**(3), 236–250 (2008)
5. Mulder, J., Serdijn, W.A., Roermund, A.H.M.: An RMS-DC converter based on the dynamic translinear principle. IEEE Solid-State Circuits **32**, 1146–1150 (1997)
6. Wey, W.S., Huang, Y.C.: A CMOS delta–sigma true RMS converter. IEEE J. Solid-State Circuits **35**(2), 248–257 (2000)
7. De La Cruz-Blas, A., Lopez-Martin, A.J., Carlosena, A., Ramirez-Angulo, J.: 1.5V current-mode CMOS true RMS–DC converter based on class-AB transconductors. IEEE Trans. Circuits Syst. II, Express Briefs **52**(7), 376–379 (2005)
8. Eskiyerli, M., Payne, A.J.: Square root domain filter design and performance. Analog Integr. Circuits Signal Process. **22**, 231–243 (2000)
9. Lopez-Martin, A.J., Carlosena, A.: A 1.5V current-mode CMOS RMS-to-DC converter. Analog Integr. Circuits Signal Process. **36**(1–2), 137–143 (2003)
10. Farshidi, E., Asiaban, H.: A new true RMS-to-DC converter using up-down translinear loop in CMOS technology. Analog Integr. Circuits Signal Process. **70**(3), 385–390 (2012)
11. Petrovic, P.: New CMOS current-mode analogue to digital power converter. Circuits, Syst. Signal Process. **36**(4), 1361–1378 (2017)
12. Alikhani, A., Ahmadi, A.: A novel current-mode four-quadrant CMOS analog multiplier/divider. Int. J. Electron. Commun. (AEÜ) **66**, 581–586 (2012)
13. Minaei, S., Sayin, O.K., Kuntman, H.: A new CMOS electronically tunable current conveyor and its application to current-mode filters. IEEE Trans. Circuits Syst.—I: Regul. Pap. **53**(7), 1448–1457 (2006)
14. Gray, P.R., Hurst, P.J., Lewis, S.H., Meyer, R.G.: Analysis and Design of Analog Integrated Circuits, 5th edn. Wiley (2009)

15. McConaghy, T., Breen, K., Dyck, J., Gupta, A.: Variation-Aware Design of Custom Integrated Circuits: A Hands-On Field Guide. Springer (2013)
16. Guide to the Expression of Uncertainty in Measurement. BIPM (2008)
17. Chaisricharoen, R., Chipipop, B., Sirinaovakul, B.: CMOS CCCII: structures, characteristics, and considerations. Int. J. Electron. Commun. (AEÜ) **64**, 540–557 (2010)
18. Baker, R.J.: CMOS Circuit Design, Layout, and Simulation, 3rd edn. Wiley (2010)
19. Serrano, G., Hasler, P.: A precision low-TC wide-range CMOS current reference. IEEE J. Solid-State Circuits **43**(2), 558–565 (2008)
20. Schmid, H.: Approximating the universal active element. IEEE Trans. Circuits Syst. II: Analog. Digit. Signal Process. **47**(11), 1160–1169 (2000)
21. Rumberg, B., Graham, D.W.: A low-power magnitude detector for analysis of transient-rich signals. IEEE J. Solid-State Circuits **47**(3), 676–685 (2012)
22. Kaewdang, K., Kumwachara, K., Surakampontorn, W.: A simple wide-band CMOS based true rms-to-dc converter. Int. J. Electron. **91**(7), 407–420 (2004)
23. Abulma'atti, M.T.: Improved analysis of implicit RMS detectors. IEEE Trans. Instrum. Meas. **58**(3), 502–505 (2009)
24. Shaterian, M., Twigg, C., Azhari, J.: MTL-based implementation of current-mode CMOS RMS-to-DC converters. Int. J. Circuit Appl. **43**(6), 793–805 (2015)
25. Kaewdang, K., Kumwachara, K., Surakampontorn, W.: A translinear-based true RMS-to-DC converter using only npn BJTs. Int. J. Electron. Commun. (AEÜ) **63**(6), 472–477 (2009)
26. Yuce, E., Minaei, S., Tokat, S.: Root-mean-square measurement of distinct voltage signals. IEEE Trans. Instrum. Meas. **56**(6), 2782–2787 (2007)
27. Farshidi, E., Sayedi, S.M.: A micropower multi decade dynamic range current-mode true RMS-to-DC converter. In: Proceedings of IEEE Northeast Workshop on Circuits and Systems (NEWCAS), pp. 1493–1496 (2007)
28. Farshidi, E., Sayedi, S.M.: A 1.2 V current-mode true RMS–DC converter based on the floating gate MOS translinear principle. Microelectron. J. **39**(2), 293–298 (2008)
29. Petrović, P., Župunski, I.: RMS detector of periodic, band-limited signals based on usage of DO-CCIIs. Measurement **46**(9), 3073–3083 (2013)

Chapter 7
Dynamic Phasors Estimation Based on Taylor-Fourier Expansion and Gram Matrix Representation

The rapid advancement of electronic technologies, including the proliferation of renewable energy sources and power devices, has brought about significant changes in the structure of traditional power systems. Micro-networks, weak networks, and island networks can experience substantial variations in electrical variables, such as the amplitude and frequency of fundamental harmonics, due to reduced short-circuit currents and harmonic injection from electronic equipment [1]. Consequently, there is a pressing need for reliable and efficient algorithms to assess and estimate the phases of electrical signals contaminated by harmonic components, with the ability to adapt to changing parameters in real-time.

In recent years, numerous algorithms have been proposed in the literature to evaluate phases in different dynamic states and controllable conditions within a network. These algorithms can generally be categorized into two groups: those based on discrete Fourier transform (DFT) methods [2–4], and those based on non-DFT methods. Each algorithm relies on a harmonic/phaser model and employs specific techniques to model the observed system parameters. Based on the modeling method and measurement approach, these algorithms can be further classified as either relying on a pure sinusoidal signal model (static model), or utilizing a non-sinusoidal model (dynamic model) [5].

Several notable algorithms have been introduced in the field. For example, a simultaneous estimation of phasor and frequency using the fast recursive Gauss–Newton algorithm was proposed in [6], while a modified Fourier transformation technique for eliminating DC offset was presented in [7]. Another method, described in [8], utilizes an adaptive band-pass filter to evaluate the phasor. The authors of [9] introduce a special angle-shifted energy operator to extract the instantaneous amplitude of the phasor. The least squares curve fitting approach was suggested in [10] to address the impact of current transformer saturation. Measurement of phase and frequency during transient processes was explored in [11], while dynamic phase estimation using maximally flat differentiators was proposed in [12], and phasorlet in [13].

P. Petrović, *Processing, Estimation and Measurement of Signals Parameters in Public Distribution Networks*, Studies in Systems, Decision and Control 502,
https://doi.org/10.1007/978-3-031-43107-4_7

Recursive wavelet transformation served as the basis for estimating phasor parameters in [14]. The Taylor Fourier algorithm [15] approximates dynamic phase using second-order Taylor expansion and a least squares observer. The Taylor Kalman method, employing a Kalman observer to achieve non-delayed dynamic phase estimation, is described in [16]. This method is further enhanced by the development of a state space approach for harmonic infiltration in [17], referred to as the Taylor Fourier Kalman method. The Prony algorithm [18] estimates the phasor using exponential amplitude and linear phase without Taylor expansion. Some methods are modifications of earlier approaches [19, 20]. The Shank method, based on least squares and consecutive delays of the unit response system, is another dynamic phasor estimation technique [21]. The TFT (Taylor-Fourier transform) harmonic analysis algorithm presented in [22] utilizes recursive multiple-resonator-based computational techniques, thereby reducing computational cost and memory requirements.

Overall, these algorithms contribute to the field of phase estimation by offering various methodologies for assessing and estimating electrical signal phases under different conditions and dynamic states.

In [23], a refined dynamic model is derived to accurately describe the complex trajectory of the dynamic phasor. This model incorporates a revised state transition equation for the rotating phasor and its derivatives. Building on this improved dynamic model [23], a modified Taylor-Kalman Filter is developed for instantaneous dynamic phasor estimation, aligning with the adaptive nature of the Kalman filter principle. In [24], an extension of the Taylor Weighted Least Squares (TWLS) algorithm is analyzed for the estimation of phasor, frequency, and ROCOF (Rate-Of-Change-Of-Frequency) parameters in electric waveforms. Reference [25] introduces a dynamic phasor estimation method for PMUs/IEDs based on a modified time-domain hybrid approach. The dynamic phasor is estimated using the Taylor expansion, with direct derivation of frequency deviation from fitting parameters to mitigate the amplification of fitting errors. Reference [26] proposes a double suboptimal-scaling factor-adaptive strong-tracking Kalman filter (DSTKF)-based algorithm for phasor measurement units under dynamic conditions, meeting the accuracy requirements of the IEEE standard C37.118.1. This method utilizes a kth Taylor polynomial to linearize the signal model's complex exponential and estimates the dynamic phasor using DSTKF. Furthermore, [27] presents a combined least square-based Prony analysis and Taylor expansion method called Taylor-Prony for dynamic phasor estimation.

In this chapter, a specific transformation using the Taylor-Fourier expansion and the corresponding Gram matrix is employed to estimate dynamic phasors from input phasorlets in the real domain, based on synchronously sampled values of the input signal [28]. The behavior of the dynamic system is described using matrix algebra associated with their Taylor-Fourier-trigonometric series representation. Explicit formulas for calculating the elements of the associated Gram matrix are provided, enhancing the computational performance of the proposed algorithm in terms of precision and speed (Namely, the proposed algorithm is based on matrix multiplication. It is a well known fact that mathematical definition of multiplication of matrices gives an algorithm that takes pqr time to multiply a matrix of type $p \times q$ with a matrix of type $q \times r$ (in the case of square $p \times p$-matrices, it takes p^3 time)). By utilizing these explicit formulas, one matrix multiplication operation is avoided, reducing the task to a series of consecutive calculations (to pr consecutive calculations) involving specific polynomials. Additionally, each element of the Gram matrix depends on the number of samples N (expressed as a polynomial in the variable N), but the time required for these calculations depends on the degree of the polynomial. Consequently, the time needed for calculating the Gram matrix remains the same even with an increased number of samples. Another advantage of employing analytical expressions for the Gram matrix elements, instead of the usual matrix multiplication interpretation, is the reduction of rounding errors that tend to accumulate during multiple computations. The explicit formulas for the elements of the Gram matrix lead to smaller round-off errors, particularly with a large number of samples.

The accuracy of the proposed method is compared to well-known methods from the literature, demonstrating its capability to track dynamic phasors, particularly when compared to other approaches based on the least squares method. Several tests, including sinusoidal and step changes in amplitude and phase, harmonic conditions, frequency tracking, and computation time, are employed to validate the proposed method based on the definition and test cases specified in the standard [29]. The potential of the proposed approach is further illustrated through MATLAB simulations of various numerical signals. This algorithm proves particularly suitable for integrating distributed generating sources with microgrids, where rapid fault detection and islanding condition identification are essential.

7.1 Dynamic Signal Model and Algorithm Description

The dynamics exhibited by a power system during oscillatory phenomena are commonly represented through the utilization of trigonometric series in modeling.

$$x(t) = \sum_{m=0}^{M} a_m(t) \sin(m\omega t + \psi_m(t)), \tag{7.1}$$

where $a_m(t)$ and $\psi_m(t)$ are the amplitude and the phase functions which represent variations of the mth dynamic harmonic over time, and $\omega = 2\pi f_1$ (f_1 is frequency of fundamental component). In an equivalent form, the behaviour of such power system can be represented as:

$$x(t) = \sum_{m=0}^{M} \{A_m(t) \sin m\omega t + B_m(t) \cos m\omega t\}, \tag{7.2}$$

where:

$$A_m(t) = a_m(t)\cos\psi_m(t)\ and\ B_m(t) = a_m(t)\sin\psi_m(t) \tag{7.3}$$

Undeniably, the following assertion is valid and holds true:

$$\psi_m(t) = \text{arctg}\frac{B_m(t)}{A_m(t)} \tag{7.4}$$

and:

$$a_m(t) = \sqrt{A_m^2(t) + B_m^2(t)} \tag{7.5}$$

Consider the approximation of variables $A_m(t)$ and $B_m(t)$ using two Taylor series expansions over a small time interval centered around the reference time $t_{ref} = 0$, denoted as follows:

$$A_m(t) = \sum_{k=0}^{K} \frac{A_m^{(k)}(0)}{k!} t^k = \sum_{k=0}^{K} A_{mk}\, t^k \tag{7.6}$$

and:

$$B_m(t) = \sum_{k=0}^{K} \frac{B_m^{(k)}(0)}{k!} t^k = \sum_{k=0}^{K} B_{mk}\, t^k \tag{7.7}$$

Consequently, we can deduce the following relationship or result:

$$x(t) = \sum_{m=0}^{M}\left\{\left(\sum_{k=0}^{K} A_{mk}\, t^k\right)\sin m\omega t + \left(\sum_{k=0}^{K} B_{mk}\, t^k\right)\cos m\omega t\right\} \tag{7.8}$$

In a discretized form, this can be expressed as a system of equations, given by:

$$x(nT) = \sum_{m=0}^{M}\left\{\left(\sum_{k=0}^{K} A_{mk}\, n^k T^k\right)\sin m\omega nT + \left(\sum_{k=0}^{K} B_{mk} n^k T^k\right)\cos m\omega nT\right\} \tag{7.9}$$

where $n = 1, 2, \ldots, N$.

For every given value of "m" ($m = 1, 2, \ldots, M$) the representation of the mth harmonic is expressed as follows:

$$x_m(nT) = \left(\sum_{k=0}^{K} A_{mk}\, n^k T^k\right)\sin m\omega nT + \left(\sum_{k=0}^{K} B_{mk} n^k T^k\right)\cos m\omega nT \tag{7.10}$$

To facilitate the estimation of unknown phasor parameters from input phasors, it is imperative to undertake specific calculations. In this regard, the processed signal samples are initially subjected to transformation using the Fourier non-recursive algorithm [30, 31]. Subsequently, these transformed samples are employed as input parameters within the algorithm presented in this research study. The mathematical formulation of the digital filtering algorithm is hereby defined as:

$$x_m^{fil}(nT) = \frac{1}{N}\sum_{k=0}^{N-1} x((k+n)T)\cos km\omega T, \quad n = 1, 2, \ldots, N \tag{7.11}$$

The filtered sample of the mth harmonic voltage or current signal at the nth index, denoted as $x_m^{fil}(n)$, can be expressed as a function of the processed signal sample $x((k+n)T)$ at the time moment $(k+n)T$. Here, T represents the sampling period, N denotes the total number of samples, and N is determined as $N = f/f_1$ (with f representing the frequency and f_1 being the frequency of the fundamental component). Upon extracting the filtered samples, forming a vector of length N, the proposed procedure is applied. To facilitate this filter transformation, the FIR structure described in [32] can be utilized. However, for the sake of simplicity and efficiency in estimating the processed phasors, the alternative structure (7.11) is selected.

Notably, by considering the case of $K = 2$, we can derive the ensuing system of equations:

$$\begin{aligned}
x_m^{fil}(T) &= \left(A_{m0} + A_{m1}T + A_{m2}T^2\right)\sin m\omega T \\
&\quad + \left(B_{m0} + B_{m1}T + B_{m2}T^2\right)\cos m\omega T, \\
x_m^{fil}(2T) &= \left(A_{m0} + A_{m1}2T + A_{m2}2^2T^2\right)\sin m\omega 2T \\
&\quad + \left(B_{m0} + B_{m1}2T + B_{m2}2^2T^2\right)\cos m\omega 2T, \\
x_m^{fil}(3T) &= \left(A_{m0} + A_{m1}3T + A_{m2}3^2T^2\right)\sin m\omega 3T \\
&\quad + \left(B_{m0} + B_{m1}3T + B_{m2}3^2T^2\right)\cos m\omega 3T, \\
&\vdots \\
x_m^{fil}(NT) &= \left(A_{m0} + A_{m1}NT + A_{m2}N^2T^2\right)\sin m\omega NT \\
&\quad + \left(B_{m0} + B_{m1}NT + B_{m2}N^2T^2\right)\cos m\omega NT.
\end{aligned} \tag{7.12}$$

The conventional algorithm employed for phasor estimation relies on the subsystem defined in Eq. (7.12). In this approach, only two column vectors are considered, resulting in the approximation of the dynamic phasor using a zero-order polynomial approximation based on Taylor's expansion. As a consequence, the estimated phasor exhibits a stair-step behavior, with varying step sizes between intervals. This modeling approach is accurate solely when the input signal remains in a stationary state. However, this is insufficient in scenarios involving power system oscillations, as the first and second derivatives become as significant as the constant term. To address this limitation, a matrix representation of the discrete system defined in Eq. (7.12) assumes the following form:

$$x_m^{fil} = H \cdot y_m \tag{7.13}$$

where:

$$x_m^{fil} = [x(T),\ x(2T),\ x(3T)\ldots, x(NT)]^T \tag{7.14}$$

$$y_m = [A_{m0},\ A_{m1}, A_{m2}, B_{m0}, B_{m1}, B_{m2}]^T \tag{7.15}$$

and:

$$H = \begin{bmatrix} \sin m\omega T & T\sin m\omega T & T^2\sin m\omega T & \cos m\omega T & T\cos m\omega T & T^2\cos m\omega T \\ \sin m\omega 2T & 2T\sin m\omega 2T & 2^2T^2\sin m\omega 2T & \cos m\omega 2T & 2T\cos m\omega 2T & 2^2T^2\cos m\omega 2T \\ \sin m\omega 3T & 3T\sin m\omega 3T & 3^2T^2\sin m\omega 3T & \cos m\omega 3T & 3T\cos m\omega 3T & 3^2T^2\cos m\omega 3T \\ \vdots & \vdots & \vdots & \vdots & \vdots & \vdots \\ \sin m\omega NT & NT\sin m\omega NT & N^2T^2\sin m\omega NT & \cos m\omega NT & NT\cos m\omega NT & N^2T^2\sin m\omega NT \end{bmatrix} \tag{7.16}$$

The objective at hand is to minimize the sum of squared residuals.

$$R(\widetilde{y}_m) = x_m^{fil} - H \cdot \widetilde{y}_m \tag{7.17}$$

Namely, our aim is to ascertain the solution to the matrix equation in order to minimize the sum of squared residuals.

$$H^T \cdot H \cdot \widetilde{y}_m = H^T \cdot x_m^{fil} \tag{7.18}$$

The matrix $H^T \cdot H$ denoted as G is commonly referred to as the Gram's matrix. In order to obtain the optimal solution in the least-squares sense, it is necessary for the Gram's matrix to be invertible, which is satisfied when the column vectors of H are linearly independent [33]. The feasibility of Gramian inversion relies on the interval size N and the order of the signal model being processed. In our approach, we assume that N is sufficiently large, such that $N > p$ (where p represents the order of the signal model). We start with assumption that $2m\omega NT = 2s\pi$, for some $s \in \mathbb{N}$ (in the case that $s = 1$, the number of samples N covers the whole period).

It is evident that the phasors can be derived from the inverse transformation of the phasorlets, and the best solution is expressed as follows:

$$\widetilde{y}_m = (H^T \cdot H)^{-1} \cdot H^T \cdot x_m^{fil} \tag{7.19}$$

The computation of the pseudo-inverse matrix $(H^T \cdot H)^{-1} \cdot H^T$ depends solely on the parameters of the chosen signal model. The estimation of the phasors at the center of the evaluation interval, where the Taylor's error is zero, is accurate when the input signal is adequately represented by the adopted model, resulting in a zero least mean squared error (LMS). Any LMS error would impact the estimation if the signal deviated from the projection subspace of the LMS algorithm.

7.2 Gram's Matrix of Dynamic Signal Model

In this section, we will investigate the Gram's matrix of the dynamic signal model (7.12), and for each element of that matrix, an explicit formula will be given.

The Gram's matrix $H^T \cdot H$ has the form.

This section focuses on examining the Gram's matrix associated with the dynamic signal model defined in Eq. (7.12). Specifically, we aim to provide explicit formulas for each element of the Gram's matrix $H^T \cdot H$. The Gram's matrix, is structured as follows:

$$H^T \cdot H = \begin{bmatrix} A & C \\ C & B \end{bmatrix} \tag{7.20}$$

where $A = \left[a_{ij}\right]_{i,j=1}^3$, $B = \left[b_{ij}\right]_{i,j=1}^3$ and $C = \left[c_{ij}\right]_{i,j=1}^3$, and for all $i, j = 1, 2, 3$ and $p = i + j - 2$, the elements a_{ij}, b_{ij} and c_{ij} are finite trigonometric series:

$$a_{ij} = \sum_{n=1}^{N} n^p \cdot \sin^2 m\omega nT \tag{7.21}$$

$$b_{ij} = \sum_{n=1}^{N} n^p \cdot \cos^2 m\omega nT \tag{7.22}$$

$$c_{ij} = \sum_{n=1}^{N} n^p \cdot \sin m\omega nT \cdot \cos m\omega nT \tag{7.23}$$

$$A_p(N, x) = \sum_{n=1}^{N} n^p \cdot \sin^2 nx \tag{7.24}$$

$$B_p(N, x) = \sum_{n=1}^{N} n^p \cdot \cos^2 nx \tag{7.25}$$

$$C_p(N, x) = \sum_{n=1}^{N} n^p \cdot \sin nx \cdot \cos nx \tag{7.26}$$

Evidently, the following relation holds true for the given condition $x = m\omega T$:

$$a_{ij} = T^p \cdot A_p(N, m\omega T),\ b_{ij} = T^p \cdot B_p(N, m\omega T) \text{ and } c_{ij} = T^p \cdot C_p(N, m\omega T) \tag{7.27}$$

The series denoted as $A_p(N, x)$, $B_p(N, x)$ and $C_p(N, x)$ can be expressed equivalently in the following manner:

$$A_p(N,x) = \frac{1}{2}\cdot\sum_{n=1}^{N} n^p - \frac{1}{2}\cdot\sum_{n=1}^{N} n^p\cdot\cos 2nx \tag{7.28}$$

$$B_p(N,x) = \frac{1}{2}\cdot\sum_{n=1}^{N} n^p + \frac{1}{2}\cdot\sum_{n=1}^{N} n^p\cdot\cos 2nx \tag{7.29}$$

$$C_p(N,x) = \frac{1}{2}\cdot\sum_{n=1}^{N} n^p\cdot\sin 2nx \tag{7.30}$$

It is widely recognized that an analytical solution for the summation of positive integer powers is well-established as:

$$S_p(N) = \sum_{n=1}^{N} n^p = \zeta(-p) - \zeta(-p, N+1) = H_n^{(-p)} \tag{7.31}$$

Here, the Riemann zeta function $\zeta(z)$ [34, 35], the Hurwitz zeta function [36], and the generalized harmonic number $H_n^{(k)}$ [37] play crucial roles. Notably, the Swiss mathematician Jacob Bernoulli (1654–1705) is credited with deriving the formula for the finite sum of powers of consecutive positive integers [38]. This formula represents the sum $S_p(N)$ as a polynomial in the variable N, with a degree $p+1$ determined by the value of the exponent. For specific values of p from the set of natural numbers $\{0, 1, \ldots, 4\}$, the corresponding polynomials are outlined in Table 7.1.

We will now direct our attention to the finite sums that involve trigonometric functions.

$$R_p(N,x) = \sum_{n=1}^{N} n^p\cdot\cos 2nx \text{ and } I_p(N,x) = \sum_{n=1}^{N} n^p\cdot\sin 2nx \tag{7.32}$$

The exponential form of the finite sums involving trigonometric functions can be expressed as follows:

Table 7.1 Sum of powers of positive integers

$S_0(N)$	N
$S_1(N)$	$\frac{1}{2}\cdot(N^2+N)$
$S_2(N)$	$\frac{1}{6}\cdot(2N^3+3N^2+N)$
$S_3(N)$	$\frac{1}{4}\cdot(N^4+2N^3+N^2)$
$S_4(N)$	$\frac{1}{30}\cdot(6N^5+15N^4+10N^3-N)$

$$E_p(N,x) = \sum_{n=1}^{N} n^p \cdot e^{\mathrm{i}2nx} = \frac{1}{(2\mathrm{i})^p} \cdot \sum_{n=1}^{N} \frac{\mathrm{d}^p}{\mathrm{d}x^p}(e^{\mathrm{i}2nx}) = \frac{1}{(2\mathrm{i})^p} \cdot \frac{\mathrm{d}^p}{\mathrm{d}x^p}\left(\sum_{n=1}^{N} e^{\mathrm{i}2nx}\right) =$$
$$= \frac{1}{(2\mathrm{i})^p} \cdot \frac{\mathrm{d}^p}{\mathrm{d}x^p}\left(e^{\mathrm{i}2x} \cdot \frac{e^{\mathrm{i}2Nx}-1}{e^{\mathrm{i}2x}-1}\right) = \frac{1}{(2\mathrm{i})^p} \cdot \frac{\mathrm{d}^p f(x)}{\mathrm{d}x^p}, \tag{7.33}$$

where:

$$f(x) = e^{\mathrm{i}2x} \cdot \frac{e^{\mathrm{i}2Nx}-1}{e^{\mathrm{i}2x}-1} \tag{7.34}$$

For arbitrary $N \in \mathbb{N}$, let functions $g_p(z) : C \to C$ be defined in the following way:

$$g_0(z) = z \cdot \frac{z^N - 1}{z-1} \; and \; g_p(z) = 2\mathrm{i} \cdot z \cdot \frac{\mathrm{d}g_{p-1}(z)}{\mathrm{d}z}, \quad p = 1, 2, \ldots \tag{7.35}$$

For $z = e^{\mathrm{i}2x}$ hold $f(x) = g_0(z)$ and:

$$\frac{\mathrm{d}^p f(x)}{\mathrm{d}x^p} = \frac{\mathrm{d}}{\mathrm{d}x}\left(\frac{\mathrm{d}^{p-1} f(x)}{\mathrm{d}x^{p-1}}\right) = \frac{\mathrm{d}}{\mathrm{d}z}\left(\frac{\mathrm{d}^{p-1} f(x)}{\mathrm{d}x^{p-1}}\right) \cdot \frac{\mathrm{d}z}{\mathrm{d}x} = \frac{\mathrm{d}}{\mathrm{d}z}(g_{p-1}(z)) \cdot \frac{\mathrm{d}z}{\mathrm{d}x} = g_p(z) \tag{7.36}$$

Thus, we obtain the following result:

$$E_p(N,x) = \frac{1}{(2\mathrm{i})^p} \cdot g_p(e^{\mathrm{i}2x}) \tag{7.37}$$

By implication, we arrive at the following conclusion:

$$R_p(N,x) = \mathrm{Re}\left\{\frac{1}{(2\mathrm{i})^p} \cdot g_p(e^{\mathrm{i}2x})\right\} \text{ and } I_p(N,x) = \mathrm{Im}\left\{\frac{1}{(2\mathrm{i})^p} \cdot g_p(e^{\mathrm{i}2x})\right\} \tag{7.38}$$

The series denoted as $A_p(N,x)$, $B_p(N,x)$ and $C_p(N,x)$ can be expressed in terms of $S_p(N)$ and $g_p(x)$ as follows:

$$A_p(N,x) = \frac{1}{2} \cdot S_p(N) - \frac{1}{2} \cdot \mathrm{Re}\left\{\frac{1}{(2\mathrm{i})^p} \cdot g_p(e^{\mathrm{i}2x})\right\} \tag{7.39}$$

$$B_p(N,x) = \frac{1}{2} \cdot S_p(N) + \frac{1}{2} \cdot \mathrm{Re}\left\{\frac{1}{(2\mathrm{i})^p} \cdot g_p(e^{\mathrm{i}2x})\right\} \tag{7.40}$$

$$C_p(N,x) = \frac{1}{2} \cdot \mathrm{Im}\left\{\frac{1}{(2\mathrm{i})^p} \cdot g_p(e^{\mathrm{i}2x})\right\} \tag{7.41}$$

Furthermore, these series can be succinctly represented in terms of $S_p(N)$, $R_p(N, x)$ and $I_p(N, x)$ using the following concise notation:

$$A_p(N, x) = \frac{1}{2} \cdot S_p(N) - \frac{1}{2} \cdot R_p(N, x) \tag{7.42}$$

$$B_p(N, x) = \frac{1}{2} \cdot S_p(N) + \frac{1}{2} \cdot R_p(N, x) \tag{7.43}$$

$$C_p(N, x) = \frac{1}{2} \cdot I_p(N, x) \tag{7.44}$$

For values of p belonging to the set of natural numbers $\{0, 1, \ldots, 4\}$, the explicit expressions of functions $g_p(z)$ obtained by p consecutive differentiations, as per formula (7.35), are presented in Table 7.2.

In the primary section of this chapter, the synchronous model is examined, wherein the value of N is selected in accordance with the relationship $Nx_0 = \pi$. By applying this condition for specific values of $p = 0, 1, \ldots, 4$ and $z_0 = e^{i2x_0}$, the resulting values of $E_p(N, x_0)$ are presented in Table 7.3.

If we represent each z_0^s in $E_p(N, z_0)$ in its expanded form as a polynomial in $\sin x_0$ and $\cos x_0$ of degree (where the degree of each term in this polynomial is the sum of the exponents in each term), $R_p(N, x_0)$ can be transformed into the form presented in Table 7.4.

In a similar manner, by expressing each z_0^s in $E_p(N, z_0)$ in its expanded form as a polynomial in $\sin x_0$ and $\cos x_0$ of degree p, $I_p(N, x_0)$ can be converted into the form illustrated in Table 7.5.

Applying formula (7.42), we derive the corresponding values of $A_p(N, x_0)$, $p = 0, 1, \ldots, 4$, which are presented in Table 7.6.

Utilizing formula (7.43), we calculate the values of $B_p(N, x_0)$, for $p = 0, 1, \ldots, 4$, as outlined in Table 7.7.

Finally, employing Eq. (7.44), we determine the values of $C_p(N, x_0)$, for $p = 0, 1, \ldots, 4$, as presented in Table 7.8.

By employing Eq. (7.27), we derive the elements of the Gram's matrix denoted as $H^T \cdot H$. Specifically, for all $i, j = 1, 2, 3$ and $p = i + j - 2$, the elements a_{ij} are expressed as follows:

$$a_{11} = \frac{N}{2}, \tag{7.45}$$

$$a_{12} = a_{21} = \frac{TN^2}{4}, \tag{7.46}$$

$$a_{13} = a_{22} = a_{31} = T^2 \cdot \left(\frac{1}{6}(N^3 - N) - \frac{1}{4} N \cdot \text{ctg}^2 m\omega T \right), \tag{7.47}$$

Table 7.2 The functions $g_p(z)$

$g_0(z)$	$\frac{z}{z-1} \cdot \left(z^N - 1\right)$
$g_1(z)$	$\frac{2\mathrm{i}z}{(z-1)^2} \cdot \left(Nz^{N+1} - (N+1)z^N + 1\right)$
$g_2(z)$	$\frac{(2\mathrm{i})^2 z}{(z-1)^3} \cdot \left(N^2 z^{N+2} - (2N^2 + 2N - 1)z^{N+1} + (N^2 + 2N + 1)z^N - z - 1\right)$
$g_3(z)$	$\frac{(2\mathrm{i})^3 z}{(z-1)^4} \cdot \begin{pmatrix} N^3 z^{N+3} - (3N^3 + 3N^2 - 3N + 1)z^{N+2} + (3N^3 + 6N^2 - 4)z^{N+1} \\ -(N+1)^3 z^N + z^2 + 4z + 1 \end{pmatrix}$
$g_4(z)$	$\frac{(2\mathrm{i})^4 z}{(z-1)^5} \cdot \begin{pmatrix} N^4 z^{N+4} - (4N^4 + 4N^3 - 6N^2 + 4N - 1)z^{N+3} \\ +(6N^4 + 12N^3 - 6N^2 - 12N + 11)z^{N+2} - (4N^4 + 12N^3 + 6N^2 - 12N - 11)z^{N+1} \\ +(N+1)^4 z^N - z^3 - 11z^2 - 11z - 1 \end{pmatrix}$

Table 7.3 The values of functions $E_p(N, x_0)$

$E_0(N, x_0)$	0	
$E_1(N, x_0)$	$\frac{q_{11} z_0^{\frac{1}{2}}}{2i \cdot \mathrm{Im} z_0^{\frac{1}{2}}}$	$q_{11} = N$
$E_2(N, x_0)$	$\frac{q_{21} z_0^{\frac{3}{2}} + q_{22} z_0^{\frac{1}{2}} + q_{23}}{\left(2i \cdot \mathrm{Im} z_0^{\frac{1}{2}}\right)^3}$	$q_{21} = N^2$ $q_{22} = -2N^2 - 2N$ $q_{33} = N^2 + 2N$
$E_3(N, x_0)$	$\frac{q_{31} z_0^{\frac{1}{2}} + q_{32} z_0 + q_{33} + q_{34} z_0^{-1}}{\left(2i \cdot \mathrm{Im} z_0^{\frac{1}{2}}\right)^4}$	$q_{31} = N^3$ $q_{32} = -3N^3 - 3N^2 + 3N$ $q_{33} = 3N^3 + 6N^2$ $q_{34} = -N^3 - 3N^2 - 3N$
$E_4(N, x_0)$	$\frac{q_{41} z_0^{\frac{5}{2}} + q_{42} z_0^{\frac{3}{2}} + q_{43} z_0^{\frac{1}{2}} + q_{44} z_0^{-\frac{1}{2}} + q_{45} z_0^{-\frac{3}{2}}}{\left(2i \cdot \mathrm{Im} z_0^{\frac{1}{2}}\right)^5}$	$q_{41} = N^4 z^{N+4}$ $q_{42} = -4N^4 - 4N^3 + 6N^2 - 4N$ $q_{43} = 6N^4 + 12N^3 - 6N^2 - 12N$ $q_{44} = -4N^4 - 12N^3 - 6N^2 + 12N$ $q_{45} = N^4 + 4N^3 + 6N^2 + 4N$

Table 7.4 The values of functions $R_p(N, x_0)$

$R_0(N, x_0)$	0
$R_1(N, x_0)$	$\frac{N}{2}$
$R_2(N, x_0)$	$\frac{1}{2} \cdot (N + N + N \cdot \mathrm{ctg}^2 x_0)$
$R_3(N, x_0)$	$\frac{1}{4} \cdot (2N^3 + 3N^2 + 3N^2 \cdot \mathrm{ctg}^2 x_0)$
$R_4(N, x_0)$	$\frac{1}{4} \cdot (2N^4 + 4N - 2N - 6N \cdot \mathrm{ctg}^4 x_0 + (4N - 8N) \cdot \mathrm{ctg}^2 x_0)$

Table 7.5 The values of functions $I_p(N, x_0)$

$I_0(N, x_0)$	0
$I_1(N, x_0)$	$-\frac{N}{2} \cdot \mathrm{ctg}\, x_0$
$I_2(N, x_0)$	$-\frac{N}{2} \cdot \mathrm{ctg}\, x_0$
$I_3(N, x_0)$	$\frac{1}{4} \cdot (3N \cdot \mathrm{ctg}^3 x_0 - (2N^3 - 3N) \cdot \mathrm{ctg}\, x_0)$
$I_4(N, x_0)$	$\frac{1}{2} \cdot (3N \cdot \mathrm{ctg}^3 x_0 - (N^4 - 3N) \cdot \mathrm{ctg}\, x_0)$

Table 7.6 The values of $A_p(N, x_0)$

$A_0(N, x_0)$	$\frac{N}{2}$
$A_1(N, x_0)$	$\frac{N^2}{4}$
$A_2(N, x_0)$	$\frac{1}{6}(N^3 - N) - \frac{1}{4}N \cdot \text{ctg}^2 x_0$
$A_3(N, x_0)$	$\frac{1}{8}(N^4 - 2N^2) - \frac{3}{8}N^2 \cdot \text{ctg}^2 x_0$
$A_4(N, x_0)$	$\frac{1}{30}(3N^5 - 10N^3 + 7N) + \frac{3}{4}N \cdot \text{ctg}^4 x_0 - \frac{1}{2}(N^3 - 2N) \cdot \text{ctg}^2 x_0$

Table 7.7 The values $B_p(N, x_0)$

$B_0(N, x_0)$	$\frac{N}{2}$
$B_1(N, x_0)$	$\frac{1}{4}(N^2 + 2N)$
$B_2(N, x_0)$	$\frac{1}{6}(N^3 + 3N^2 + 2N) + \frac{1}{4}N \cdot \text{ctg}^2 x_0$
$B_3(N, x_0)$	$\frac{1}{8}(N^4 + 4N^3 + 4N^2) + \frac{3}{8}N^2 \cdot \text{ctg}^2 x_0$
$B_4(N, x_0)$	$\frac{1}{30}(3N^5 + 15N^4 + 20N^2 - 8N) - \frac{3}{4}N \cdot \text{ctg}^4 x_0 + \frac{1}{2}(N^3 - 2N) \cdot \text{ctg}^2 x_0$

Table 7.8 The values $C_p(N, x_0)$

$C_0(N, x_0)$	0
$C_1(N, x_0)$	$-\frac{N}{4} \cdot \text{ctg}\, x_0$
$C_2(N, x_0)$	$-\frac{N}{4} \cdot \text{ctg}\, x_0$
$C_3(N, x_0)$	$\frac{1}{8} \cdot (3N \cdot \text{ctg}^3 x_0 - (2N^3 - 3N) \cdot \text{ctg}\, x_0)$
$C_4(N, x_0)$	$\frac{1}{4} \cdot (3N \cdot \text{ctg}^3 x_0 - (N^4 - 3N) \cdot \text{ctg}\, x_0)$

$$a_{23} = a_{32} = T^3 \cdot \left(\frac{1}{8}(N^4 - 2N^2) - \frac{3}{8}N^2 \cdot \text{ctg}^2 m\omega T\right), \tag{7.48}$$

$$a_{33} = T^4 \cdot \left(\frac{1}{30}(3N^5 - 10N^3 + 7N) + \frac{3}{4}N \cdot \text{ctg}^4 m\omega T - \frac{1}{2}(N^3 - 2N) \cdot \text{ctg}^2 m\omega T\right). \tag{7.49}$$

Furthermore, for all $i, j = 1, 2, 3$ and $p = i + j - 2$, the elements b_{ij} can be determined as follows:

$$b_{11} = \frac{N}{2}, \tag{7.50}$$

$$b_{12} = b_{21} = \frac{T}{4} \cdot (N^2 + 2N), \tag{7.51}$$

$$b_{13} = b_{22} = b_{31} = T^2 \cdot \left(\frac{1}{6}(N^3 + 3N^2 + 2N) + \frac{1}{4}N \cdot \text{ctg}^2 m\omega T\right), \tag{7.52}$$

$$b_{23} = b_{32} = T^3 \cdot \left(\frac{1}{8}(N^4 + 4N^3 + 4N^2) + \frac{3}{8}N^2 \cdot \mathrm{ctg}^2 m\omega T \right), \tag{7.53}$$

$$b_{33} = T^4 \cdot \left(\frac{1}{30}(3N^5 + 15N^4 + 20N^2 - 8N) - \frac{3}{4}N \cdot \mathrm{ctg}^4 x \right.$$
$$\left. + \frac{1}{2}(N^3 - 2N) \cdot \mathrm{ctg}^2 m\omega T \right). \tag{7.54}$$

In conclusion, the elements c_{ij} for all $i, j = 1, 2, 3$ and $p = i + j - 2$ can be obtained as follows:

$$c_{11} = 0, \tag{7.55}$$

$$c_{12} = c_{21} = -\frac{TN}{4} \cdot \mathrm{ctg}\, m\omega T, \tag{7.56}$$

$$c_{13} = c_{22} = c_{31} = -\frac{T^2 N}{4} \cdot \mathrm{ctg}\, m\omega T, \tag{7.57}$$

$$c_{23} = c_{32} = \frac{T^3}{8} \cdot \left(3N \cdot \mathrm{ctg}^3 x_0 - (2N^3 - 3N) \cdot \mathrm{ctg}\, m\omega T\right), \tag{7.58}$$

$$c_{33} = \frac{T^4}{4} \cdot \left(3N \cdot \mathrm{ctg}^3 x_0 - (N^4 - 3N) \cdot \mathrm{ctg}\, m\omega T\right) \tag{7.59}$$

The presented form of the Gram's matrix representation demonstrates that the computation of the pseudo-inverse $(H^T \cdot H)^{-1} \cdot H^T$ can be executed with enhanced speed and precision.

7.3 Simulation Results

The proposed phasor measurement scheme is utilized to estimate the coefficients of the power signal in nonstationary scenarios. In this context, we compared the results obtained from the proposed algorithm with those of other established techniques. Through simulation tests, we assessed the algorithms' capability to track changes in harmonics. Furthermore, we conducted tests to evaluate the performance of the proposed technique in compliance with the conditions outlined in the Standard [29] for synchronized phasor measurement systems in power systems. The initial verification of the proposed estimation procedure was conducted using a test signal specifically designed to replicate oscillating conditions.

$$x(t) = a(t)\cos(2\pi f_1 t + \phi(t)), \tag{7.60}$$

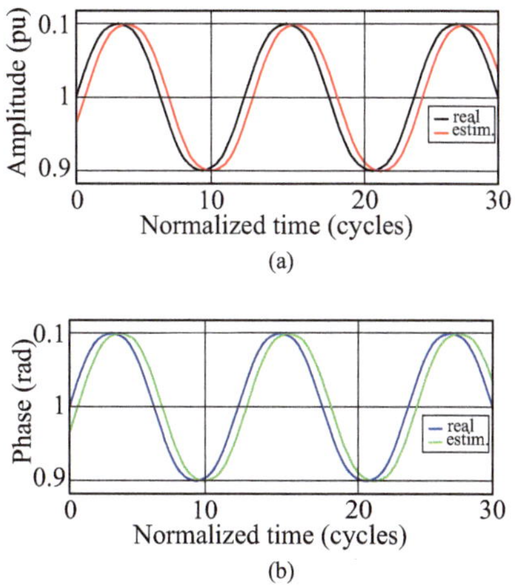

Fig. 7.1 **a** Amplitude estimation, **b** phase estimation

where $a(t) = a_0 + a_1 \sin 2\pi f_a t$, $\phi(t) = \phi_0 + \phi_1 \sin 2\pi f_\phi t$, $a_0 = \phi_0 = 1$, $a_1 = \phi_1 = 0.1$, $f_a = f_\phi = 5$ and $N = 24$.

From the aforementioned relationships, it is evident that the test signal is sampled at a frequency of 1200 Hz, resulting in 24 samples within a 20 ms duration window, which corresponds to the period of the signal with a fundamental frequency of 50 Hz. Figure 7.1a, b illustrate the estimated amplitude and phase of the fundamental dynamic phasor using the proposed estimation method. Notably, a noticeable distinction between the proposed method and other techniques based on least square methods [15, 21, 27], as well as methods employing Kalman filters [16, 17, 23, 32], lies in the occurrence of estimation delay attributable to the utilization of the data window.

The Total Vector Error (TVE) is a well-established criterion for evaluating the accuracy of phasor estimation. It allows us to quantify the error in the estimation of phasor magnitude and angle by defining it as:

$$TVE = \frac{|X_r - X_e|}{|X_r|} \tag{7.61}$$

The Total Vector Error (TVE) is computed by comparing the real value X_r and the estimated value X_e. In Fig. 7.2, the TVE is depicted for the proposed method over the first ten cycles, with a compensation of one cycle time delay in the estimated phasor, as described in [27]. It can be inferred that the observed error is primarily due to the introduced delay, akin to the findings in [15, 21]. Notably, due to the modification in the calculation of phasor parameters, the magnitude of this error is smaller compared to other methods based on the least squares approach. Furthermore, it becomes comparable to [16, 17, 23, 39], while significantly smaller than [27, 40–42]. The proposed approach incorporates the estimation of phasor derivatives, including phasor speed and acceleration, leading to a reduction in estimation error.

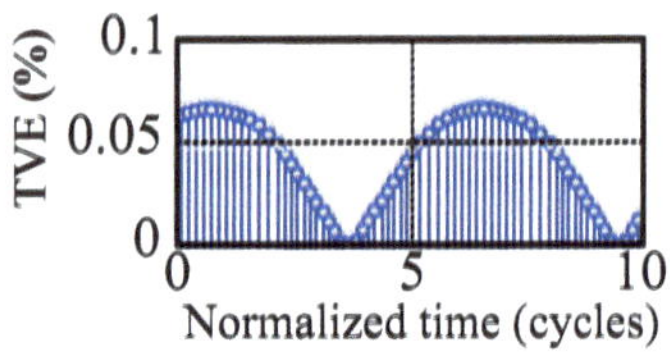

Fig. 7.2 Total Vector Error (TVE)

Moreover, it enables the computation of the frequency of the processed signal and facilitates fault detection and power swing detection.

7.3.1 Amplitude Oscillation Case

The capability of the proposed approach to accurately track an arbitrary oscillating amplitude is demonstrated in the following example, where:

$$x(t) = a(t)\sum_{h=0}^{\infty} c_h \cos 2\pi h f_1 t \quad (7.62)$$

In the example provided, the input signal to the estimator is represented by $x(t)$, with $f_1 = 50$ Hz and non-zero Fourier coefficients ($c_1 = 1$, $c_3 = 0.4$ and $c_5 = 0.2$). Additionally, the amplitude oscillation, denoted as $a(t)$, is characterized by a second-order polynomial expressed as:

$$a(t) = at^2 + bt + c \quad (7.63)$$

With coefficients $a = -4$, $b = 4$ and $c = 0$, the corresponding digital signal is sampled at a rate of 24 samples per cycle (sampling frequency of 1200 Hz). The oscillation of the main-original signal and its corresponding estimated signal components are visually depicted in Fig. 7.3. The proposed estimation method effectively captures the dynamics of all processed input signal components. However, it is evident that the estimation quality diminishes as the number of harmonics increases. This reduction in accuracy can be attributed to the relatively small and fixed number of samples, which cannot provide precise reconstruction for higher harmonic components. If the number of samples were adaptively adjusted based on the harmonic order, the estimation accuracy would be comparable to that achieved for the fundamental harmonic component. Moreover, the non-ideal characteristics of the filter function pose an additional challenge.

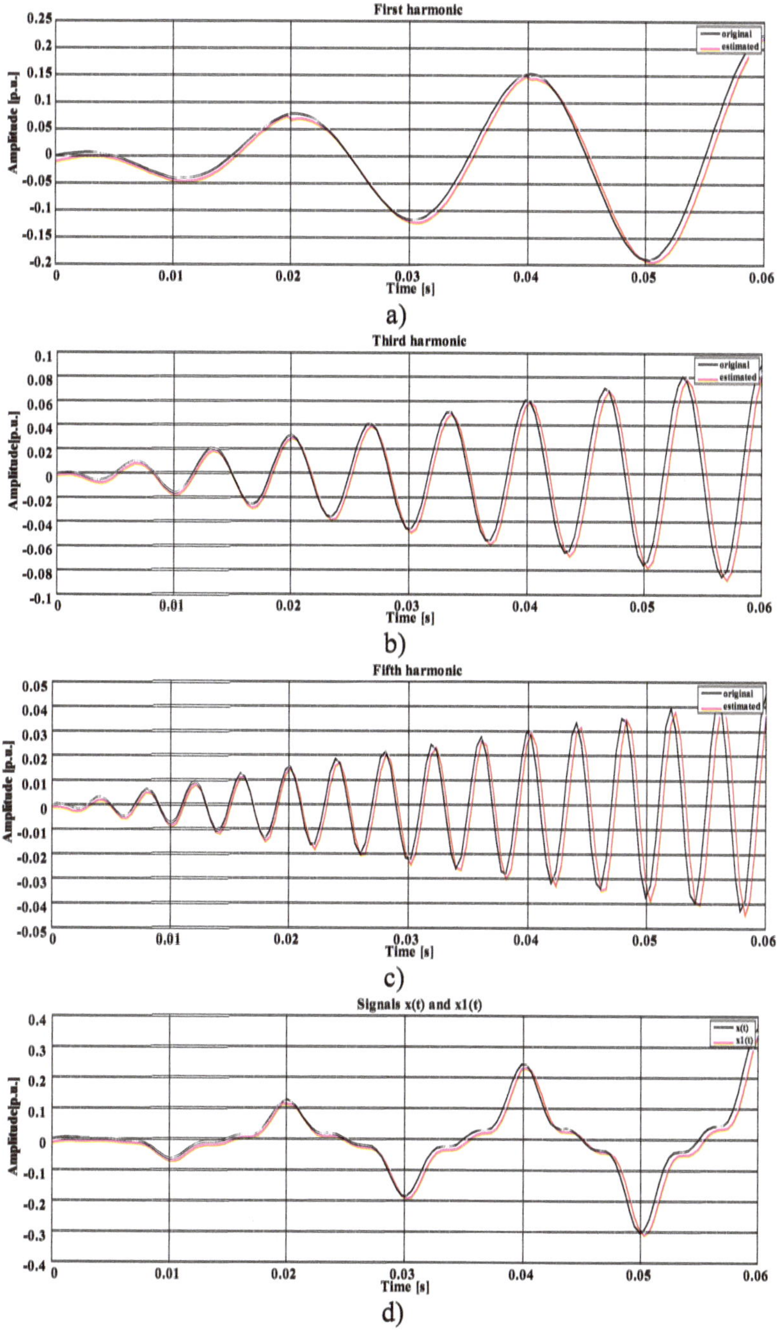

Fig. 7.3 Measured and estimated waveforms in the case of a time-varying signal (7.62), **a** tracking the first phasors, **b** tracking the 3rd phasors, **c** tracking the 5th phasors, **d** measured and estimated waveforms of main input signal

7.3.2 Magnitude-Phase Step Test

In this section, we conduct amplitude and phase step tests utilizing the proposed estimation procedure. The test signal is defined as specified in [29], where the step changes in amplitude and phase occur independently.

$$x(t) = X_m(1 + K_a u(t - t_s))cos(\omega t + k_\varphi u(t - t_s)) \tag{7.64}$$

where $u(t)$ represents the unit step function, k_a is the amplitude step size, k_φ is the phase step size, and t_s is the step instant. In this specific test, a signal with a phase angle step of 10° and an amplitude step change from 0.9 to 1 p.u. in the fundamental component is applied to the algorithm. The chosen parameters are as follows: $X_m = \sqrt{2}, \omega = 2\pi \cdot 50\,\text{Hz}, k_a = +0.1\,\text{p.u.}, k_\phi = \pi/18$, while the measurement noise is assumed to be negligible. The results obtained from the amplitude and phase step tests are presented in Fig. 7.4. The upper subplot displays the transient response of the amplitude or the phase, while the lower subplot illustrates the Total Vector Error (TVE).

The proposed approach exhibits relatively long transient periods but with minimal overshoots, as observed in Fig. 7.4. Based on these results, it is evident that the proposed method accurately estimates the amplitude and phase once the transition period has elapsed. This is because the proposed approach incorporates modifications to Tailor's-Fourier expansion, making it more suitable for smooth amplitude and phase changes. Furthermore, the duration of this transition period in the proposed approach is shorter compared to certain other methods starting from the same point [17, 27], which can be attributed to the distribution of poles in the transfer function within the unit circle in the z-plane (with greater pole spacing). Moreover, the response does not exhibit significant overshoot, unlike [16, 17, 23], which can be attributed to the amplitude and phase discontinuities associated with step changes and the absence of prior assessment using a sample window.

Table 7.9 lists the performance indices defined in the standard [29], including response time, delay time, and overshoot, for both the amplitude and phase step change tests. The proposed algorithm achieves performance levels that meet the M-Class measurement requirements outlined in the standard.

7.3.3 Frequency Step Test

In this simulation test, a signal with a 5 Hz frequency step is employed to assess the transient response in a frequency step scenario. Although a 5 Hz frequency step is unlikely to occur in a power grid, this test is designed in accordance with the guidelines outlined in the IEEE standard [29].

Fig. 7.4 Estimation of **a** amplitude, **b** phase, and **c** TVE in percentage

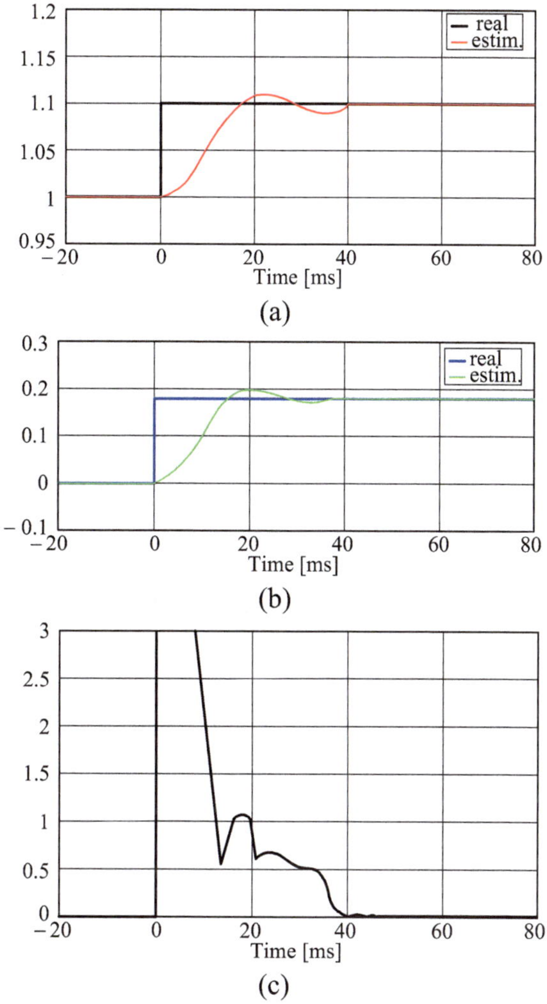

Table 7.9 Performance indices for phasor estimation under step changes in amplitude and phase using the proposed estimation algorithm

	Response time (ms)	Delay time (ms)	Overshoot
Amplitude step	19.7	8.86	0.008 (p.u.)
Phase step		9.67	0.02 (rad)

$$\begin{aligned} x(n) &= \cos 2\pi f_1 n\Delta t, && 0 < n < 10N_1, \\ x(n) &= \cos 2\pi (f_1 + 5) n\Delta t, && 10N_1 \le n. \end{aligned} \tag{7.65}$$

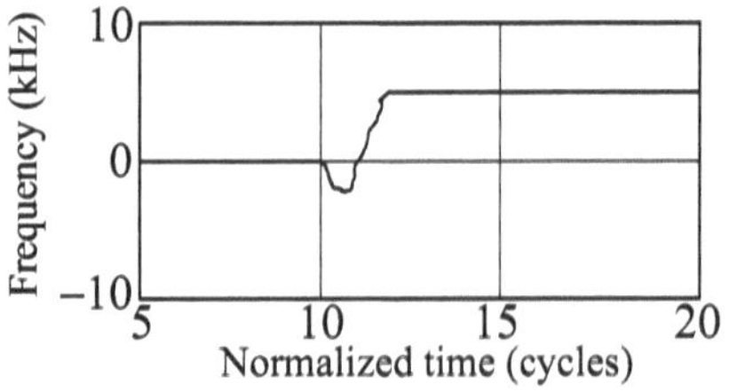

Fig. 7.5 Frequency estimation during frequency step test

The frequency was determined by computing the ratio of roots of the characteristic equation using the extracted unknown coefficients A_{m0}, A_{m1}, A_{m2}, B_{m0}, B_{m1}, B_{m2} from Eq. (7.12) and 2π, following the approach outlined in [15, 27]. The proposed method exhibits similar characteristics and evaluation procedures to the previous test scenario involving magnitude-phase step conditions. Figure 7.5 demonstrates that the proposed processing method effectively tracks a frequency step of +5 Hz, in contrast to the traditional approach employed for phasor processing.

7.3.4 *Harmonic Infiltration Test*

The input signal, defined by Eq. (7.66), was employed to validate the effectiveness of the proposed methods in the presence of harmonic conditions.

$$\begin{aligned} &x(t) = a(t)\cos(\omega_1 t + \phi(t)) + 0.05cos5\omega_1 t + 0.03cos7\omega_1 t,\\ &a(t) = a_0 + a_1 \sin 2\pi f_a t, \quad \phi(t) = \phi_0 + \phi_1 \sin 2\pi f_\phi t,\\ &a_0 = 1,\ \phi_0 = 0.5,\ a_1 = 0.1,\ \phi_1 = 0.05,\ f_a = f_\phi = 5,\ N = 24. \end{aligned} \tag{7.66}$$

Figure 7.6 depicts the output of the proposed method in the presence of harmonics. The fundamental phasor estimation based on the proposed modification exhibits a notable absence of harmonics, highlighting the superiority of the harmonic modification in this paper. In this particular test, N is set to 24, allowing for the estimation of dynamic phasors for the first $N - 1 = 23$ harmonics. To expand the range of harmonics that can be accurately estimated, the number of samples per cycle should be increased.

7.3.5 *Frequency Response*

To assess the frequency characteristics of the proposed estimator, we define its transfer function in the following form:

$$H(z) = \frac{\hat{p}(z)}{X(z)} \tag{7.67}$$

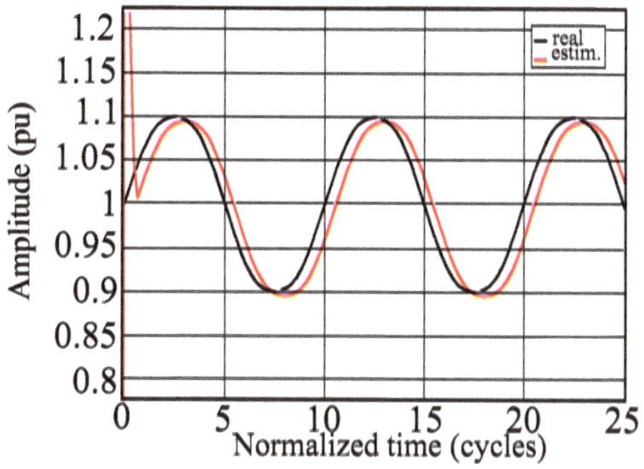

Fig. 7.6 Amplitude estimation in harmonic conditions

The transfer function of the proposed estimator is represented by the equation, where $X(z)$ and $\hat{p}(z)$ denotes the z-transform of the input (main signal) and output (estimated dynamic phasor). The frequency response of the proposed method is illustrated in Fig. 7.7. The magnitude responses exhibit asymmetry, which is expected as it pertains to complex filters. In the vicinity of the fundamental frequency, complete elimination of the negative frequency and complete pass of the positive frequency are observed. The flatter transfer characteristic results in lower distortion in phasor estimation and provides zero-gain for non-fundamental components. Consequently, the harmonics at the output are effectively eliminated, in contrast to the approaches presented in [16, 21, 23].

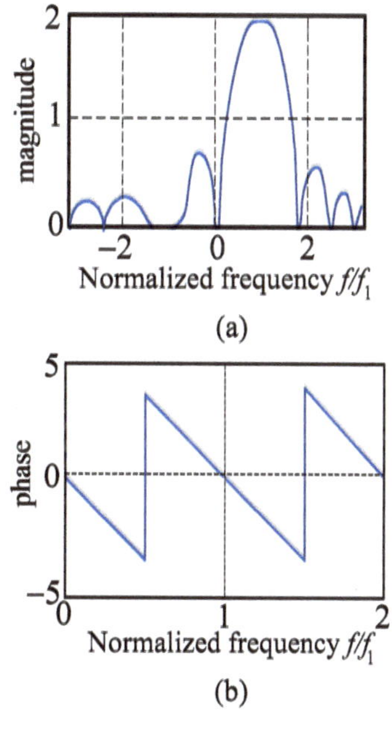

Fig. 7.7 Frequency response **a** magnitude response, **b** phase response

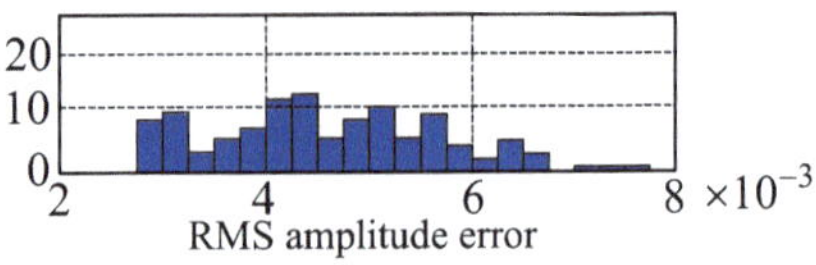

Fig. 7.8 RMS error of amplitude estimation

7.3.6 Error Bounds

To establish the boundaries of potential errors in phasor estimation, signals with variable envelopes were utilized.

$$a(t) = 1 + 0.5 \sum_{i=1}^{3} e^{\frac{t}{\tau_i}} \cos 2\pi f_1 t \tag{7.68}$$

The time constants (τ_i) were generated using a uniform random process within the range of [20, 40] cycles. Similarly, the three frequencies were randomly generated within the intervals of (1, 3), (3, 5) and (5, 7) Hz. The error is computed using the following formula:

$$rms - error = \sqrt{\sum (a(n) - \hat{a}(n))} \tag{7.69}$$

In the equation above, $a(n)$ and $\hat{a}(n)$ represent the real and estimated amplitudes, respectively. Figure 7.8 displays histograms illustrating the errors obtained by the proposed method.

The Root Mean Square (RMS) error for the proposed estimation method falls within the range of 3×10^{-3} and 8×10^{-3}. This error is attributed to the omission of the third derivative term in the Taylor expansion during the phasor estimation process. Notably, this error is significantly lower compared to the estimated errors reported in [43–48].

7.3.7 Noise, Sub-harmonics and Inter-harmonics Infiltration

Analyzing the estimation error of the proposed method under varying noise levels proves beneficial. For this evaluation, the transient monitor (TM) criterion, as introduced in [21], is employed. TM serves as an indicator to monitor abrupt changes in the measured signal and is computed using the following formula:

$$[t_n] = \left[S_n - \hat{S}_n\right]$$
$$TM = \sum_{n=r-N_1}^{n=r} |t_n| \tag{7.70}$$

The *TM* criterion is computed using the formula, where $\hat{S}_n$ represents the real (measured) value of the sample, and t_n is the estimation error obtained by comparing the recomputed sample (obtained using the estimated dynamic phasor) with the real value. *TM* serves as a quality measure for phasor estimation. In Fig. 7.9, the *TM* is plotted as a function of noise variance for the proposed method as well as other well-known phasor estimation methods. As the noise variance increases, the errors also increase, resulting in an upward trend. It can be observed from Fig. 7.9 that both the least squares-based methods and our proposed method exhibit a stable trend before reaching a critical point with a variance of approximately 10^{-2}. Beyond this point, the transient monitor increases rapidly due to the growing error variance. Generally, the Kalman-based methods [16, 17, 23] show lower *TM* values compared to the least squares-based methods. However, our proposed algorithm outperforms other Taylor expansion-based methods when the noise variance is low, as depicted in Fig. 7.9. Moreover, it can be concluded that the proposed method demonstrates a downward trend with increasing sampling frequency. This indicates that the estimation accuracy improves with a higher sampling rate, contrary to the findings in [21] where increased noise sensitivity was observed at high sampling rates.

The behavior of estimation methods in the presence of interharmonic and subharmonic components, which are commonly found in practical installations where these measurements are applied, is not explicitly defined in the standard [29]. Interharmonics, in particular, can be challenging to detect and isolate, especially if they are

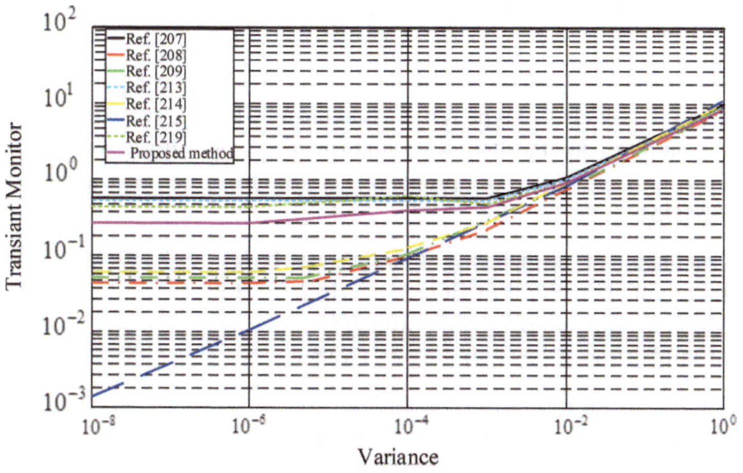

Fig. 7.9 Transient monitor versus noise variance

located within the frequency band of interest of the phasor dynamic model. To further evaluate the performance of the phasor estimation method described in this paper, additional components were introduced to the signal. Specifically, a subharmonic frequency of 20 Hz and two interharmonic components at frequencies of 57.5 and 85 Hz [5] were added to a sinusoidal signal containing the aforementioned contents. An out-of-band interference test was devised to assess the ability to filter interfering signals that could be aliased into the measurement. It is important to note that this test is only applicable to the M class and is excluded for the P class [40].

The accuracy of the proposed algorithms in phasor estimation is closely linked to their filtering capabilities. When the interharmonic component falls within the passband of the algorithm, which is designed to capture the dynamic phenomena of interest, the Total Vector Error (TVE) is approximately 3% for interharmonic components at 57.5 Hz. In cases where interharmonics are located in the stopband frequencies, the TVE depends on the stopband attenuation. For example, for an interharmonic frequency of 85 Hz, the TVE is approximately 0.5%. Similar results are obtained for the subharmonic component at 20 Hz. These findings demonstrate improved performance compared to those reported in [5, 40].

Furthermore, the proposed approach for phasor estimation was tested in a scenario where the input signal contained a third harmonic with an amplitude set at 20% of the fundamental component, along with an interharmonic frequency of 75 Hz. This level of harmonic content is commonly observed in current waveforms in control applications such as current tracking in shunt active power filters (APFs) in marine networks or smart distributed grids with renewable energy sources [40]. The TVE in the estimation of the third harmonics in this test was approximately 1%. In all the aforementioned tests, the number of samples used was $N = 24$.

7.3.8 Computational Complexity and Simulation Time

The algorithm takes two matrices as inputs: matrix H, which is of type $N \times 2(K+1)$, and matrix x_m^{fll}, which is of type $N \times 1$, as defined by Eqs. (7.16) and (7.13) respectively. The output of the algorithm is a column matrix $\widetilde{y}_m$, which is of type $2(K+1)\times 1$, obtained using Eq. (7.19). The proposed algorithm can be decomposed into the following steps:

(A1) Calculation of the elements of Gram's matrix $H^T \cdot H$ by (7.45)–(7.59).
(A2) Inversion $(H^T \cdot H)^{-1}$ of Gram's matrix $H^T \cdot H$.
(A3) Multiplication of $(H^T \cdot H)^{-1}$ with H^T.
(A4) Multiplication of $(H^T \cdot H)^{-1} \cdot H^T$ with x_m^{fll}.

(A1) By utilizing formulas (7.45)–(7.59) for the computation of the elements of Gram's matrix $H^T \cdot H$, the task can be simplified to $4(K+1)^2$ performing a sequence of specific value calculations based on polynomials. Let c_0 represent the maximum computational time required for executing (7.45)–(7.59). In this case, the computational time for $H^T \cdot H$ does not exceed $4(K+1)^2 c_0$.

(A2) The solution to Eq. (7.19) involves the inversion of the Gram's matrix. To obtain the inverse of the $2(K+1) \times 2(K+1)$-matrix $H^T \cdot H$, the Lower–Upper (LU) decomposition method can be employed, requiring $\frac{4}{3} \cdot (2(K+1))^3 + \frac{3}{2} \cdot (2(K+1))^2 - \frac{5}{6} \cdot 2(K+1)$ floating-point operations (FLOPs).

(A3) In this step, the multiplication of the matrix $(H^T \cdot H)^{-1}$, of type $2(K+1) \times 2(K+1)$, by the matrix H^T, of type $2(K+1) \times N$, requires a computational complexity of $4(K+1)^2 N$ FLOPs.

(A4) In this step, the multiplication of the matrix $(H^T \cdot H)^{-1} \cdot H^T$, of type $2(K+1) \times N$, by the matrix x_m^{fil}, of type $N \times 1$, requires $2(K+1)N$ FLOPs.

Based on all the aforementioned procedures and steps, the computational complexity of the proposed algorithm does not exceed:

$$\begin{aligned} &4(K+1)^2 c_0 + \frac{4}{3} \cdot (2(K+1))^3 + \frac{3}{2} \cdot (2(K+1))^2 - \frac{5}{6} \cdot 2(K+1) \\ &\quad +4(K+1)^2 N + 2(K+1)N = \left(4K^2 + 8K + 4\right)c_0 \\ &+\left(\frac{32}{3}K^3 + 38K^2 + \frac{127}{3}K + 15\right) + \left(4K^2 + 10K + 6\right)N. \end{aligned} \tag{7.71}$$

operations-FLOPs, which practically defined its computational costs. If the step (A1) is replaced with:

(A′1) Multiplication of H^T with H,

The computational time of this step increases compared to the time needed for performing step (A1). Since matrix H is an $N \times 2(K+1)$ matrix, the resulting matrix $H^T \cdot H$ is an $2(K+1) \times 2(K+1)$ matrix, and multiplication of matrices gives an algorithm that takes time $4(K+1)^2 N$ to multiply the matrix H^T of type $2(K+1) \times N$ with the matrix H of type $N \times 2K$. This time complexity depends on the number of samples N. On the other hand, the computational time in step (A1) is independent of N, allowing for efficient processing of a large number of samples without compromising performance. Additionally, if the time c_0 is the maximum computational time required for executing steps (7.45)–(7.59), it specifically corresponds to the time needed for implementing step (7.54). If $\operatorname{ctg} x$ is regarded as $\operatorname{ctg} x \approx {}^1/_x - {}^x/_3$, then its computational time takes 3 FLOPs for basic arithmetic operations, the computational time for step (7.54) can be approximated by a finite number c.

To practically determine the computation time required for estimating unknown phasor parameters using the proposed method in real conditions, the frequency of the processed signal is increased. This allows for an assessment of whether the method is suitable for offline or online processing. A comparison was conducted with well-known methods in the literature, and the results are presented in Fig. 7.10. The hardware used for these comparisons consisted of an Intel(R) Core(TM) i5-4460 processor operating at 3.2 GHz with 16 GB of RAM. The proposed modification in the algorithm demonstrated significantly reduced computation time compared to

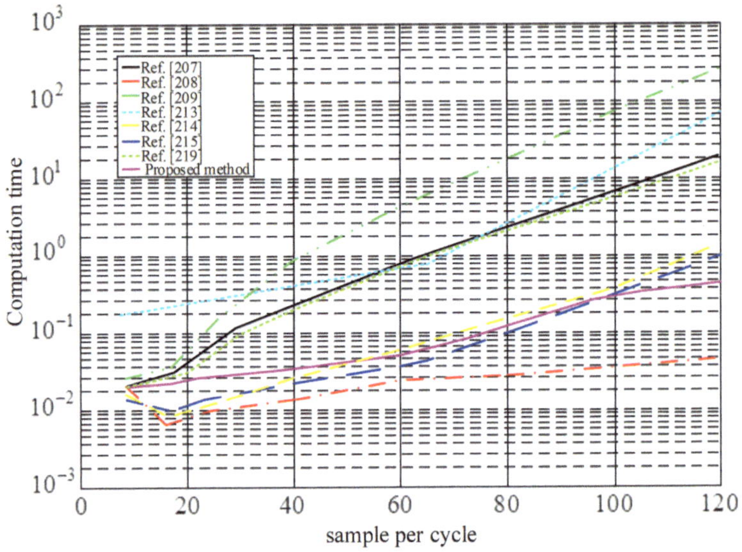

Fig. 7.10 Computation time versus sampling frequency

other least squares-based methods [24, 27], approaching the speed of methods such as [21, 23, 40–42].

In this section of the monograph, we have utilized the dynamic phasor concept to estimate the variable amplitude and phase in the processing of oscillating signals in modern power grids. By introducing a specific modification to the Taylor-Fourier expansion and the corresponding Gram matrix, we have developed a more precise and computationally efficient algorithm for estimating all the unknown parameters of the processed signal, in accordance with the defined conditions in the standard. Through extensive simulations, we have demonstrated the accuracy and capability of the proposed method in terms of Total Vector Error (TVE), harmonic conditions, frequency tracking, and computational time.

The proposed method has been thoroughly investigated under various conditions and has proven to be a valuable and efficient tool for detecting signal components under dynamic conditions. The simulation results have shown that the proposed technique provides accurate estimates and has the ability to track phase, frequency, and amplitude changes in nonstationary signals. The specific form of the proposed algorithm allows for reduced computational cost and memory requirements, making it highly suitable for real-time digital systems.

References

1. Rohten, A.J., et al.: Enhanced predictive control for a wide time variant frequency environment. IEEE Trans. Ind. Electron. **63**(9), 5827–5837 (2016)
2. Kamwa, I., Samantaray, S.R., Joos, G.: Wide frequency range adaptive phasor and frequency PMU algorithm. IEEE Trans. Smart Grid **5**(2), 569–579 (2014)
3. Chung, S.-K.: Harmonic power flow determination using the fast Fourier transform. IEEE Trans. Power Deliv. **2**(2), 530–535 (1991)
4. Marques, G.A.G., Ribeiro, M.V., Duque, C.A., Ribeiro, P.F., da Silva, E.A.B.: A controlled filtering method for estimating harmonics of off-nominal frequencies. IEEE Trans. Smart Grid **3**(1), 38–49 (2012)
5. Castello, P., Lixia, M., Muscas, C., Pegoraro, P.A.: Impact of the model on the accuracy of synchrophasor measurement. IEEE Trans. Instrum. Meas. **61**(8), 2179–2188 (2012)
6. Dash, P., Krishnanand, K., Patnaik, R.: Dynamic phasor and frequency estimation of timevarying power system signals. Int. J. Electr. Power Energy Syst. **44**(1), 971–980 (2013)
7. Kang, S.-H., Lee, D.-G., Nam, S.-R., et al.: Fourier transform-based modified phasor estimation method immune to the effect of the DC offsets. IEEE Trans. Power Deliv. **24**(3), 1104–1111 (2009)
8. Kamwa, I., Pradhan, A.K., Joos, G.: Adaptive phasor and frequency-tracking schemes for wide-area protection and control. IEEE Trans. Power Deliv. **26**(2), 744–753 (2011)
9. Jin, X., Wang, F., Wang, Z.: A dynamic phasor estimation algorithm based on angle-shifted energy operator. Sci. China Technol. Sci. **56**(6), 1322–1329 (2013)
10. Lee, D.-G., Kang, S.-H., Nam, S.-R.: Phasor estimation algorithm based on the least square technique during CT saturation. J. Electr. Eng. Technol. **6**(4), 459–465 (2011)
11. Phadke, A.G., Kasztenny, B.: Synchronized phasor and frequency measurement under transient conditions. IEEE Trans. Power Deliv. **24**(1), 89–95 (2009)
12. Platas-Garza, M.A., Platas-Garza, J., de la O Serna, J.A.: Dynamic phasor and frequency estimates through maximally flat differentiators. IEEE Trans. Instrum. Meas. **59**(7), 1803–1811 (2010)
13. de la O Serna, J.A.: Phasor estimation from phasorlets. IEEE Trans. Instrum. Meas. **54**(1), 134–143 (2005)
14. Ren, J., Kezunovic, M.: Real-time power system frequency and phasors estimation using recursive wavelet transform. IEEE Trans. Power Deliv. **26**(3), 1392–1402 (2011)
15. de la O Serna, J.A.: Dynamic phasor estimates for power system oscillations. IEEE Trans. Instrum. Meas. **56**(5), 1648–1657 (2007)
16. de la O Serna, J.A., Rodriguez-Maldonado, J.: Instantaneous oscillating phasor estimates with Taylor-Kalman filters. IEEE Trans. Power Syst. **26**(4), 2336–2344 (2011)
17. de la O Serna, J.A., Rodŕiguez-Maldonado, J.: Taylor–Kalman–Fourier filters for instantaneous oscillating phasor and harmonic estimates. IEEE Trans. Instrum. Meas. **61**(4), 941–951 (2012)
18. de la O Serna, J.A.: Synchrophasor estimation using Prony's method. IEEE Trans. Instrum. Meas. **62**(8), 2119–2128 (2013)
19. de la O Serna, J.A., Martin, K.E.: Improving phasor measurements under power system oscillations. IEEE Trans. Power Syst. **18**(1), 160–166 (2003)
20. de la O Serna, J.A.: Reducing the error in phasor estimates from phasorlets in fault voltage and current signals. IEEE Trans. Instrum. Meas. **56**(3), 856–866 (2007)
21. Munoz, A.T., de la O Serna, J.A.: 'Shanks' method for dynamic phasor estimation. IEEE Trans. Instrum. Meas. **57**(4), 813–819 (2008)
22. Kušljevic, M.D., Tomic, J.J.: Multiple-resonator-based power system Taylor-Fourier harmonic analysis. IEEE Trans. Instrum. Meas. **64**(2), 554–563 (2015)
23. Liu, J., Ni, F., Tang, J., Ponci, F., Monti, A.: A modified Taylor-Kalman filter for instantaneous dynamic phasor estimation. In: 2012 3rd IEEE PES Innovative Smart Grid Technologies Europe (ISGT Europe), pp. 1–7, 14–17 Oct 2012

24. Belega, D., Fontanelli, D., Petri, D.: Dynamic phasor and frequency measurements by an improved Taylor weighted least squares algorithm. IEEE Trans. Instrum. Meas. **64**(8), 2165–2178 (2015)
25. Qian, C., Kezunovic, M.: Dynamic synchrophasor estimation with modified hybrid method. In: 2016 IEEE Power & Energy Society Innovative Smart Grid Technologies Conference (ISGT), pp. 1–5, 6–9 Sept 2016
26. Huang, C., Xie, X., Jiang, H.: Dynamic phasor estimation through DSTKF under transient conditions. IEEE Trans. Instrum. Meas. **66**(11), 2929–2936 (2017)
27. Khodaparast, J., Khederzadeh, M.: Dynamic synchrophasor estimation by Taylor-Prony method in harmonic and nonharmonic conditions. IET Gener. Transm. Distrib. **11**(18), 4406–4413 (2017)
28. Petrovic, P.B., Damljanovic, N.: Dynamic phasors estimation based on Taylor-Fourier expansion and Gram matrix representation. Math. Probl. Eng. **2018**, 7613814 (2018)
29. IEEE standard for synchrophasor measurements for power systems. IEEE Std C37.118.1-2011 (Revision of IEEE Std C37.118-2005), pp. 1–61 (2011)
30. Petrovic, P., Rozgic, D.: Computational effective modified Newton-Raphson algorithm for power harmonics parameters estimation. IET Signal Proc. **12**(5), 590–598 (2018)
31. Rechka, S., Ngandui, E., Xu, J., Sicard, P.: Performance evaluation of harmonics detection methods applied to harmonics compensation in presence of common power quality problems. Math. Comput. Simul. **63**(3–5), 363–375 (2003)
32. Petrovic, P.: Algorithm for simultaneous parameters estimation of multi-harmonic signal. Metrol. Meas. Syst. **19**(4), 693–702 (2012)
33. Ben-Israel, A., Greville, T.N.E.: Generalized Inverses: Theory and Applications, 2nd edn. Springer, New York, NY, USA (2003)
34. Apostol, T.M.: Introduction to Analytic Number Theory. Springer-Verlag, New York-Heidelberg-Berlin (1976)
35. Ivic, A.: The Riemann Zeta Function. Wiley (1985)
36. Dwilewicz, R., Minac, J.: The Hurwitz zeta function as a convergent series. Rocky Mt. J. Math. **36**(4), 1191–1219 (2006)
37. Knuth, D.: The Art of Computer Programming, Volume 1: "Fundamental Algorithms", 3rd edn. Addison-Wesley (1997)
38. Dwilewicz, R., Minac, J.: Values of the Riemann zeta function at integers. Mater. Math. **2009**(6), 1–26 (2009)
39. Khodaparast, J., Khederzadeh, M.: Least square and Kalman based methods for dynamic phasor estimation: a review. Prot. Control. Mod. Power Syst. **2**(1), 1–18 (2017)
40. Kušljevic, M.D., Tomic, J.J., Poljak, P.D.: Maximally flat-frequency-response multiple-resonator-based harmonic analysis. IEEE Trans. Instrum. Meas. **66**(12), 3387–3398 (2017)
41. de la O Serna, J.A.: Synchrophasor measurement with polynomial phaselocked-loop Taylor–Fourier filters. IEEE Trans. Instrum. Meas. **64**(2), 328–337 (2015)
42. Karimi-Ghartemani, M., Mojiri, M., Bakhshai, A.,et al.: A phasor measurement algorithm based on phase-locked loop. In: 2012 IEEE PES Transmission and Distribution Conf. and Exposition (T&D), pp. 1–6. IEEE (2012)
43. Langella, R., Testa, A.: The effects of integration intervals on recursive RMS value and power measurements in nonsinusoidal conditions. IEEE Trans. Instrum. Meas. **60**(9), 3047–3057 (2011)
44. Vizireanu, D.N., Halunga, S.V.: Analytical formula for three points sinusoidal signals amplitude estimation errors. Int. J. Electron. **99**(1), 149–151 (2012)
45. Petrović, P., Župunski, I.: A new time-domain method for calculation of basic electric values in public distribution networks. Trans. Inst. Meas. Control **35**(3), 279–288 (2013)
46. Petrovic, P.: Modified formula for calculation of active power and RMS value of band-limited AC signals. IET Sci. Meas. Technol. **6**(6), 510–518 (2012)

47. IEC 50160: Voltage characteristics of electricity supplied by public distribution networks (2011)
48. Martens, O., Trampark, H., Liimets, A., Nobel, P., Veskimester, A., Jarvalt, A.: DSP-based power quality monitoring device. In: IEEE International Symposium on Intelligent Signal Processing, pp.1–5 (2008)

MIX
Papier aus verantwortungsvollen Quellen
Paper from responsible sources
FSC® C105338

If you have any concerns about our products,
you can contact us on
ProductSafety@springernature.com

In case Publisher is established outside the EU,
the EU authorized representative is:
Springer Nature Customer Service Center GmbH
Europaplatz 3, 69115 Heidelberg, Germany

Printed by Libri Plureos GmbH
in Hamburg, Germany